Sy

Systems-thinking for Safety

A short introduction to the theory
and practice of systems-thinking

Simon Bennett

PETER LANG
Oxford · Bern · Berlin · Bruxelles · New York · Wien

Bibliographic information published by Die Deutsche Nationalbibliothek.
Die Deutsche Nationalbibliothek lists this publication in the Deutsche National-
bibliografie; detailed bibliographic data is available on the Internet at
http://dnb.d-nb.de.

A catalogue record for this book is available from the British Library.

Library of Congress Cataloging-in-Publication Data:
Names: Bennett, Simon, 1958- author.
Title: Systems-thinking for safety : a short introduction to the theory and
   practice of systems-thinking / Simon Bennett.
Description: Oxford ; New York : Peter Lang, [2018] | Series: Systems
   thinking for safety. | Includes bibliographical references and index.
Identifiers: LCCN 2018058175 | ISBN 9781788743778 (alk. paper)
Subjects: LCSH: Accident investigation. | System theory. | Risk management. |
   Disasters--Prevention.
Classification: LCC HD7262.25 .B46 2018 | DDC 363.12001/1--dc23 LC record
available at https://lccn.loc.gov/2018058175

Cover design by Peter Lang Ltd.

ISSN 1661-6863
ISBN 978-1-78874-377-8 (print) • ISBN 978-1-78874-700-4 (ePDF)
ISBN 978-1-78874-701-1 (ePub) • ISBN 978-1-78874-702-8 (mobi)

© Peter Lang AG 2019

Published by Peter Lang Ltd, International Academic Publishers,
52 St Giles, Oxford, OX1 3LU, United Kingdom
oxford@peterlang.com, www.peterlang.com

Simon Bennett has asserted his right under the Copyright, Designs and Patents Act,
1988, to be identified as Author of this Work.

All rights reserved.
All parts of this publication are protected by copyright.
Any utilisation outside the strict limits of the copyright law, without the
permission of the publisher, is forbidden and liable to prosecution.
This applies in particular to reproductions, translations, microfilming,
and storage and processing in electronic retrieval systems.

This publication has been peer reviewed.

Printed in Germany

For Mrs Nancy Doreen Bennett
Captain Nigel 'Tommy' Box

Contents

List of figures	ix
List of tables	xi
Preface	xiii
Acknowledgements	xv
Introduction	1
CHAPTER 1 Systems-thinking	5
CHAPTER 2 Systems-thinking in practice	19
CHAPTER 3 A case study in systems-thinking	99
Conclusions	121
Glossary of terms	125
Bibliography	133
Index	147

Figures

Figure 1.	The network space: how systems theory conceptualises socio-technical systems	10
Figure 2.	Actor-network theory (ANT) conceptualises a socio-technical system as a *hybrid-collectif* of stories and things	11
Figure 3.	Almost 80 per cent of China's coal mines are unregulated. Entrance to a small mine	15
Figure 4.	Dryden: the causal soup (not an exhaustive list of actants)	24
Figure 5.	In happier times: Nimrod XV230 at the 2005 Waddington Air Show, England	25
Figure 6.	RAF Nimrod XV230 loss: the causal soup (not an exhaustive list of actants)	28
Figure 7.	Possible depths of analyses in a systems-thinking-informed investigation	30
Figure 8.	The USS *John S. McCain*	31
Figure 9.	A severely damaged USS *John S. McCain* limping towards Changi Naval Base in the Republic of Singapore	34
Figure 10.	Helicopter mission over the inundation caused by the tsunami	45
Figure 11.	Windscale – site of Britain's most serious nuclear accident	47
Figure 12.	Mindlessness in respect of design and operation increases system vulnerability	53
Figure 13.	Mindfulness is positively linked to safety	56
Figure 14.	International Atomic Energy Agency experts walk the Fukushima Daiichi site post-disaster	58
Figure 15.	Rigs in the Cromarty gas field photographed in untypical weather. Turbulence in the Middle East in	

	the 1960s and 1970s incentivised the rapid exploitation of local oil and gas reserves, even in the unforgiving North Sea	61
Figure 16.	In the same way that gas-guzzlers were a product of the era of cheap oil, the Piper Alpha disaster was in part a product of the oil-shocks of the 1970s, Reaganomics and Thatcherism	69
Figure 17.	Statistics pertaining to death and injury should be considered against a range of factors, including the scale, complexity and adverse operating conditions of the North Sea oil and gas industry. Rigs and pipelines require constant maintenance	74
Figure 18.	The rig sank thirty-six hours after the initial explosion	78
Figure 19.	The Line Operations Safety Audit virtuous circle, from resourcing the audit to organisational change	93
Figure 20.	Adverse events have complex origins	99
Figure 21.	An obscenity in the world's fifth largest economy?	100
Figure 22.	The charred hulk of Grenfell Tower photographed nearly eleven months after the fire	104
Figure 23.	Edwards's innovative SHEL(L) model of aviation as a socio-technical system	131
Figure 24.	In the Cold War motion-picture classic *Fail-Safe*, a malfunction in a tightly coupled, high-speed command-and-control computer sees a fleet of Convair B-58s tasked to destroy Moscow	132

Tables

Table 1.	Elements of a positive safety culture	71
Table 2.	How a LOSA is performed	92
Table 3.	LOSA's TOCs	94

Preface

In a paper titled *Advancing socio-technical systems thinking: A call for bravery*, Davis, Challenger, Jayewardene and Clegg (2014: 171) observed: 'Whilst [socio-technical systems thinking] has made an impact, we argue that we need to be braver, encouraging the approach to evolve and extend its reach. In particular, we need to: extend our conceptualization of what constitutes a system [and] apply our thinking to a much wider range of complex problems'. This is my contribution to Davis, Challenger, Jayewardene and Clegg's call-to-arms.

Acknowledgements

The author would like to thank everyone from the world of aviation who has helped him over the last twenty years. Aviation has changed the world for the better. Long may it thrive.

Introduction

This monograph is a manifesto for the systems-thinking-informed approach to incident and accident investigation. It is in tune with the mission of the *Systems-thinking for Safety* series – to publish books on systems-thinking that are entertaining, engaging and catholic in their appeal.

The series takes in a wide audience, from sixth-form, high-school and university students to journalists and the informed general reader (the 'woman on the Clapham omnibus'). Academic concepts are explained.

The monograph demonstrates the universal applicability of the systems-thinking-informed approach, from marine accident investigation to understanding the social, economic and political origins of a deadly tower block fire.

Systems-thinking for Safety: A short introduction to the theory and practice of systems-thinking is unoriginal and original. Unoriginal because it stands on the shoulders of giants – Elwyn Edwards, Barry Turner, Bruno Latour, John Law, Steven Woolgar, Charles Perrow, Erik Hollnagel, Diane Vaughan, James Reason and other luminaries of the world of letters. Original because it is populist. The Peter Lang series *Systems-thinking for Safety* provides a mechanism for popularising the work of thinkers. This will help make the world a safer place.

The monograph commences with a discussion about the origins and nature of the systems-thinking approach to risk assessment and management. The discussion includes a review of actor-network theory (ANT), a powerful but under-used device for explaining how the world works.

To demonstrate that systems-thinking can be a force for good and to engage the reader, the bulk of the monograph is given over to case studies of systems-thinking in action, specifically its contribution to the investigation of incidents, accidents and near-misses in complex socio-technical systems such as aviation, maritime operations, nuclear power generation and oil and gas extraction. The monograph is expansive. It aims to engage as wide an audience as possible.

Featured socio-technical systems are represented in various ways: sometimes with a simple list of actants (the human and non-human elements of a socio-technical system); sometimes with an actant-map; sometimes with a topographical diagram. Whatever the representational device, the intention is always to convey the *complex origins* of incident, accident and near-miss.

The aphorism 'Things are never quite as simple as they seem' is especially true in the field of incident and accident investigation. Often, the obvious answer is the wrong answer. Frequently, the obvious answer has more to do with political convenience than with justice.

On 6 February 1958, at Munich Airport, a British European Airways (BEA) Airspeed Ambassador (Flight 609) crashed on take-off, killing twenty-three and injuring nineteen. The dead and injured included players from Manchester United Association Football Club. The German authorities attributed the Ambassador's poor runway performance to wing ice. The British government, determined to rebuild relations with West Germany, accepted the German government's verdict. Politically, crew error was a convenient narrative: 'Recently declassified British files show that, while the authorities privately took Thain's [James Thain captained Flight 609] side all along, they did not exert more pressure in order to avoid embarrassing the Germans in the fraught postwar atmosphere' (Leroux 2008).

It was subsequently found that the aircraft's failure to achieve take-off speed was caused not by wing ice but by runway contamination (slush). In the 1950s, the impact of runway contamination on take-off performance was poorly understood. A broken Thain had to wait until 1969 – a decade after Munich – to be exonerated.

Bennett (2010) argues that the British government, by colluding with the German authorities, exposed passengers and flight-crew to risk: 'The politically-inspired scapegoating of Captain Thain ... served no constructive purpose. Indeed, because it obscured one of the underlying causes of the disaster (the impact of runway contamination on take-off rolls), it subsequently exposed passengers and employees to extra risk'.

The botched investigation into the 1958 Munich air disaster demonstrates:

Introduction

- that facts can be manipulated to support a political end (in this case, a perceived need to build relations with a former enemy);
- that, whatever the emotional, reputational or financial cost, the interests of individual citizens may be subordinated to those of the state;
- that the obvious answer is not necessarily the correct answer;
- that the mechanisms of malfunction and error are complex and often poorly understood despite scientific advances;
- that finding the truth takes commitment and effort (without his family's vocal campaign, in all probability Captain Thain would not have been exonerated). Getting to the truth is *effortful*;
- that, as American author John Steinbeck suggested, the public can be unthinking and cruel. In an interview, Thain's daughter Sebuda recalled: 'I was bullied and tormented at school over dad's involvement with Munich. It was very upsetting and difficult to avoid' (Thain cited in *Manchester Evening News* 2008).

Potentially, the systems-thinking-informed approach to risk management is an antidote to injustice and a means of limiting human suffering. On the night of 6 July 1988, Piper Alpha, an oil-and-gas production platform, was eviscerated in a series of massive explosions and conflagrations. One hundred and sixty-seven were killed. This monograph argues that Piper was operated in a way that invited incident and accident.

Systems-thinking could have safeguarded the platform. Its absence made Piper vulnerable. On 6 July 1988, that vulnerability manifested in terror and suffering. As a worker who witnessed the destruction of Piper Alpha recalled: 'I could see through the smoke these men ... standing on the heli-deck. They were waving their arms. ... But no-one could get to them. It took 20 minutes before the whole thing blew. I saw men climbing down the legs, jumping into the water. The heat was unbearable. There was men with their faces burned off. ... [We] could do nothing except stand around like sheep as those men died' (McBain cited in Hall 1995: 50). Those who went to help faced danger: 'The heat was so bad that I felt I just had to find cover, even though we must have been 100 yards away from the thing. I saw a rope hanging over the trawler's side, so I grabbed hold and just jumped. ... The skipper just opened up both engines ... and I was dragged underneath the surface. ... We can't thank those boats enough. I saw one inflatable dinghy pay the price ... it had its tanks ignited by the heat, and exploded ... leaving only a melted stretch of rubber on the sea' (Punchard cited in Hall 1995: 49).

As demonstrated by Piper Alpha, Texas City, Macondo, Challenger, Chernobyl, Fukushima, Grenfell Tower and numerous other disasters, the consequences of not rectifying systemic weaknesses through the lifetime of a system – from design to refurbishment to decommissioning – can be lethal.

Features

The monograph includes a glossary of terms to aid comprehension. The glossary provides accessible definitions of terms such as practical drift, safety migration, passive learning, active learning and other terms in common usage in the field of risk management. The monograph's main arguments are summarised in eight boxes.

CHAPTER 1

Systems-thinking

Provenance

Promoted by Professors Elwyn Edwards (1972), Barry Turner (1978), Charles Perrow (1983, 1984), James Reason (1990), Diane Vaughan (1996), Erik Hollnagel (2004) and Sidney Dekker (2014b), systems-thinking first registered in the public consciousness in a serious way in the early 1990s (Maurino, Reason, Johnston and Lee 1998), with the publication of the Honourable Mr Justice Virgil P. Moshansky's (1992) systems-thinking-informed investigation into the 1989 Dryden air disaster – where, in the context of operational pressures and resource issues, a crew's failure to de-ice their aircraft cost twenty-four lives.

Moshansky's innovation was that he considered both the immediate and proximate causes of the disaster, including the politics and day-to-day management of Canada's air transportation sector. Moshansky's high-fidelity, inclusive investigation led him to conclude that Dryden 'was the result of a failure in the air transportation system' (Moshansky 1992: 5–6). Instead of identifying a single 'probable cause', Moshansky's report made 191 recommendations (from which a list of active and latent failures could be deduced), including some pertaining to the evolving structure of Canada's recently deregulated air transportation industry: 'The report … criticised deregulation of the aircraft industry, which started in Canada in 1985. … It found that even though safety controls were supposed to have been maintained, deregulation brought an increase in paperwork and fewer officials overseeing safety requirements' (Farnsworth 1992). Moshansky's report was system-thinking's 'big bang' event.

With reference to Edwards's (1972) conceptualisation of the air transport industry as a socio-technical system and Reason's (1990, 1997, 2013) work on failure mechanisms, Dryden was a 'system accident' that originated in the spaces and interactions between the aviation system's components, including:

- personnel
- equipment
- training regimes
- rules
- laws
- free-market ideology
- shareholder and investor agendas
- political agendas.

According to Maurino, Reason, Johnston and Lee, the publication in 1992 of Moshansky's *Commission of Inquiry* report marked 'the dawning of a new age'. The *Final Report*, containing nearly 200 recommendations, institutionalised 'the organisational approach to accident investigation' (Maurino, Reason, Johnston and Lee 1998: 84). This was no small achievement given the obstacles confronting Judge Moshansky. As he recalled in 1995: 'During the early stages of the Inquiry, counsel for the regulator attempted to limit the scope of the Inquiry with threats to limit my mandate by seeking an order in the Federal Court of Canada. When it became clear that intimidation would not succeed, these attempts were abandoned' (Moshansky cited in Maurino, Reason, Johnston and Lee 1998: vii). Personal and intellectual courage are useful attributes for those seeking to overturn bad regimes.

Contemporary manifestation

Regarding the question of the origins of incident, accident and near-miss, systems theory rejects person-centric explanations in favour of system-centric explanations. Systems theory explains failure in terms of system characteristics such as:

Systems-thinking

- interactive complexity
- coupling
- production pressures
- employee churn
- reactive patching
- practical drift
- safety migration
- normalisation of deviance
- non-linear interactions
- emergent behaviours
- intractability
- regulatory capture.

Systems-thinking has been nurtured over many years by numerous thinkers, including Henry Fairlie (1955), Professor Elwyn Edwards (1972), Professor Barry Turner (1978), Professor Michelle Callon (1981), Professor Bruno Latour (1981), Professor Charles Perrow (1983, 1984), Professor Stephen Woolgar (1986), Professor James Reason (1990), Professor John Law (1991), Professor Dietrich Dörner (1996), Professor Diane Vaughan (1996), Professor Erik Hollnagel (2004), Professor Christopher Johnson (2005), Dr Richard Holden (2009), Professor Kent D. Miller (2009) and Professor Sidney Dekker (2014b).

While the thinkers listed here focused on a variety of phenomena, from the stratagems employed by natural scientists to secure research funding or ensure the hegemony of a particular theory, to the latent causes of failure in high-technology artefacts such as production platforms, nuclear power stations and passenger aircraft, a common theme runs through their work – that understanding the *why* or *how* of something requires an understanding of the system or context. Specifically, it requires an understanding of the social, economic and political fabric and of how that fabric is manipulated to serve an end. Putting it another way: understanding the general is the *sine qua non* of understanding the particular.

One of the (unintentional) pioneers of the systems approach was English political commentator Henry Fairlie, whose celebrated 1955 Political Commentary article in *The Spectator* examined how, and in whose interests, power was exercised in Britain. Fairlie argued first, that power was exercised in the context of a network of informal, elite social relations. These he referred to collectively as The Establishment. Fairlie went on to argue that this network of informal, elite social relations influenced *how* power was exercised. Specifically, he argued that power was frequently exercised in the interests of a privileged few:

> The exercise of power in Britain (more specifically, in England) cannot be understood unless it is recognised that it is exercised socially. Anyone who has at any point been close to the exercise of power will know what I mean when I say that the 'Establishment' can be seen at work in the activities of, not only the Prime Minister, the Archbishop of Canterbury and the Earl Marshal, but of such lesser mortals as the chairman of the Arts Council, the Director-General of the BBC, and even the editor of *The Times Literary Supplement* ... Somewhere near the heart of the pattern of social relationships which so powerfully controls the exercise of power in this country is the Foreign Office. By its traditions and its methods of recruitment, the Foreign Office makes it inevitable that the members of the Foreign Service will be men (and the Foreign Service is one of the bastions of masculine English society) who, to use a phrase which has been used a lot in the past few days, 'know all the right people'. (Fairlie 1955: 5)

Because system topography changes over time, systems theory focuses on the system-as-found (rather than the system-as-designed). It focuses on the *lived reality* of production systems (such as power generation, minerals extraction, air service provision, health-care delivery, events-management or housing provision).

Non-linear interactions, emergence and other system phenomena render system behaviour unpredictable. Put simply, it is difficult – if not impossible – to know how a complex system will perform under all circumstances. Complex socio-technical systems may respond differently to stressors (endogenous and exogenous) over time: sometimes because different personnel respond differently to the same challenge; sometimes because the same personnel choose a different solution; sometimes because no solution is applied.

Most often, incidents, accidents and near-misses originate in a mélange of factors, both technical and social (Turner 1994; International Civil Aviation Organisation 1995; Hollnagel 2004; Dekker 2006; Harris 2014). The Transportation Safety Board of Canada (2014: 10) observes: '[A]n accident is never caused by just one factor'. Harris (2014: 91) observes: '[A]ccidents seldom have a solitary cause. ... Error is the product of design, procedures, training and/or the environment, including the organisational environment'. The ICAO (1995: 21) observes: 'In all unsafe acts it is possible to identify numerous contributory situational and task factors, such as poor communications, time-pressure, inadequate tools and equipment, poor procedures and instructions and inadequate training. Personal factors such as preoccupation, distraction, false perceptions, incomplete or inaccurate knowledge and misperception of hazards are also readily identifiable. However, flawed organisational processes and latent organisational failures are the source of most unsafe acts committed by operational personnel'. The Single European Sky ATM Research unit (SESAR) notes: '[The] systems perspective ... suggests the human is rarely, if ever, the sole cause of an accident. The systems perspective considers a variety of contextual and task-related factors that interact with the human operator within the aviation system to affect operator performance' (Single European Sky ATM Research unit 2018).

Sometimes foretold (Lagadec 1982), malfunction and failure *emerge* from a system's *network space*, conceptualised by Challenger, Clegg and Robinson (2009: 90) as comprising goals, people, buildings/infrastructure, technology, culture and processes/procedures (see Figure 1).

Actor-network theory

Systems theory informs, and is informed by, actor-network theory (ANT). Actor-network theory (Callon and Latour 1981; Callon and Law 1997) conceives of socio-technical systems as purposive assemblies of stories and things. Together, these stories and things (actants) form a proactive network or *hybrid-collectif* (see Figure 2).

Each element – whether animate or inanimate – possesses agency (actor-network theory's Principle of Generalised Symmetry). That is, a capacity to act upon the environment to further the network's interests. Network actants seek to influence (translate) other actants. There is a positive correlation between the efficacy and durability of an actor-network and

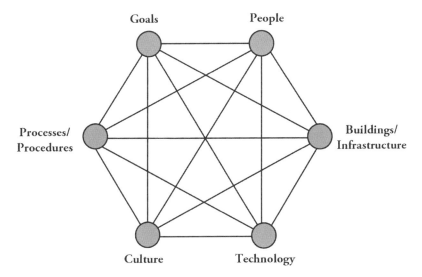

Figure 1. The network space: how systems theory conceptualises socio-technical systems (Challenger, Clegg and Robinson 2009: 90).

the complimentarity (alignment) of its actants. The more closely aligned the actants, the stronger and more effective the actor-network. A football team provides a useful analogy. The better the players and manager work together, the more likely it is that the team will win matches. The more each player's skill-set complements (supports) the skill sets of the other players, the more likely it is that the team will win matches. Complimentarity creates synergy. It makes the whole greater than the sum-of-parts.

Translation and alignment are ongoing processes: 'Stability and social order … are continually negotiated as a social process of aligning interests' (Monteiro 2012: 5). To endure, an actor-network must be alert to threats. Restlessness is one of its survival mechanisms.

Actor-networks (socio-technical systems such as agriculture or power generation or health-care) exist within larger systems. Levels interact: the micro level acts upon the meso and macro levels. The meso level upon the micro and macro levels. The macro level upon the meso and micro levels.

The man-made world is increasingly populated by systems-of-systems – meta systems such as the global financial system that is made up of national actors such as pension funds, stock exchanges, commodity exchanges,

Systems-thinking

Figure 2. Actor-network theory (ANT) conceptualises a socio-technical system as a *hybrid-collectif* of stories and things.

private banks, national banks and governments, supra-national actors such as the World Bank and International Monetary Fund (IMF), frameworks such as the North American Free Trade Agreement (NAFTA) and Organisation of the Petroleum Exporting Countries (OPEC), and regional political actors such as the African Union, Arab League, European Union (EU), Commonwealth of Independent States (CIS) and Organisation of American States (OAS). The global economy consists of myriad, mutually affecting actor-networks. Meta systems display many of the characteristics of their component systems (for example, heterogeneity and emergent behaviours) (Maier 1998; De Laurentis 2005). In the argot of political science, nation-states are subject to 'multi-level governance' by 'quasi supranational bodies' (Cerny 2010).

Actor-network theory provides us with a means of analysing and explaining every type of social phenomenon, from a major disaster such as Piper Alpha to the exposing of corruption within the New York Police Department (NYPD).

In the 1960s and 1970s, corruption amongst police officers was a feature of the NYPD's culture (Burnham 1970; Maas 1973; Kilgannon 2010). Many officers took bribes or were members of cabals that syphoned off money from extortion or drugs rackets. In 1967, one officer, Frank Serpico, decided to act. He spoke to senior officers and officials at City Hall. He supplied them with detailed accounts of racketeering and police corruption. No action was taken. The corrupt practices continued: 'Officer Serpico was appalled at the cliquishness and the payoffs – free meals as well as big, blatant bribes – from criminals, gamblers, numbers men and ordinary merchants whom he saw as a beat cop in Brooklyn's 81st Precinct and later while working vice and racketeering. [Serpico] refused to accept such grease, and became despised for it both inside and outside the department' (Kilgannon 2010).

Frank Serpico understood that in going to the authorities he had put his life in jeopardy. And so it came to pass. During a drugs bust, Serpico was shot in the face. Some believe he was set up by corrupt police officers: '[Serpico's] long and loud complaining about widespread corruption in the New York Police Department made him a pariah on the force. The patrolman [was] shot in the face during a 1971 drug bust while screaming for backup from his fellow officers, who then failed to immediately call for an ambulance' (Kilgannon 2010).

In the argot of actor-network theory, in the 1960s, corrupt police practices spawned an actor-network-of-corruption that extended from the street to police headquarters and City Hall. A network that, through processes of heterogeneous engineering, translated police officers (of all ranks), public officials, and the institutions they served. The actor-network-of-corruption included the infrastructure of the NYPD – the police vehicles, weapons, cameras, radios, station-houses and other facilities that were used in the service of corruption. It fomented a perverse organisational culture that framed corrupt officers as moral, and honest officers as immoral.

Until officer Serpico, no one had resisted translation. At least, no one had done so in a way that challenged the actor-network-of-corruption's hegemony. The manner in which Serpico resisted translation was telling. Having failed to recruit (that is, translate) senior officers and City Hall officials to his anti-corruption crusade, Serpico, with the help of fellow officer David Durk, approached the *New York Times* (*NYT*) – a city institution yet to be translated by the NYPD's actor-network-of-corruption. By approaching the NYT, Serpico and Durk engineered a counter-network, that, in theory, had the power and will to challenge the *status quo*.

Initially, Serpico and Durk's anti-corruption actor-network consisted of themselves, the *New York Times's* journalists, the owners, the readership, the newspaper's infrastructure (the offices, presses, delivery vans, street-corner vendors, advertising hoardings, etc.), ideals such as freedom of the press, truth, justice and the rule of law and, of course, the newspaper itself, a transient, flimsy artefact composed of paper and ink that was re-made every day.

The strategy worked. On 25 April 1970, the *NYT*'s David Burnham published a front-page article titled *Graft Paid to Police Here Said to Run Into Millions*. In the argot of ANT, the article bent space around itself. It accrued political and moral capital. Possessing agency, Burnham's article 'pressured Mayor John V. Lindsay to form the Knapp Commission, before which Mr Serpico testified that "the atmosphere does not yet exist in which an honest police officer can act without fear of ridicule or reprisal from fellow officers"' (Kilgannon 2010). The Knapp Commission became a key actant in New York's anti-corruption actor-network.

Burnham's (1970) uncompromising polemic helped bend space around his article and helped imbue it with agency:

> Narcotics dealers, gamblers and businessmen make illicit payments of millions of dollars a year to the policemen of New York, according to policemen, law-enforcement experts and New Yorkers who make such payments themselves. Despite such widespread corruption, officials in both the Lindsay administration and the Police Department have failed to investigate a number of cases of corruption brought to their attention, sources within the department say … The policemen and private citizens who talked to *The Times* describe a situation in which payoffs by gamblers

to policemen are almost commonplace, in which some policemen accept bribes from narcotics dealers, in which businessmen throughout the city are subjected to extortion to cover up infractions of law, and in which internal payoffs among policemen seem to have become institutionalised.

Even an artefact as transient and fragile as a newspaper can act on the world. Newspapers possess agency.

Actor-network theory is versatile. It can help us understand police corruption in the NYPD. It can help us understand why China's deep-mining industry is so lethal. Compared to the safety record of deep-mines in first-world countries like the United States, the safety record of China's deep-mines is poor (Homer 2009; Ming-Xiao, Tao, Miao-Rong, Bin and Ming-Qiu 2011; MacLeod 2014). Worryingly, it is poor when compared to the safety record of deep-mines in developing nations: '[China's] mine-safety record is significantly worse than that of other large producers who are similarly underdeveloped' (Homer 2009: 424). In 2013, 1,049 miners either died or were reported missing in China's deep mines (MacLeod 2014). In the space of a week in December 2016, sixty miners died in three separate mining disasters (Associated Press 2016). Reasons for the poor safety record include:

- the disposition of the deep-mining industry. The dispersed nature of the industry makes oversight difficult (see Figure 3). Remoteness blankets violations;
- high coal prices. When prices rise, profit-seeking owners overstress plant and labour, creating latent errors/resident pathogens;
- the absence of trade-union organising. Chinese society is at the wrong end of the pluralism spectrum;
- a workforce desperate for gainful employment. Desperate people are inclined to turn a blind eye to safety breaches: '[U]nskilled and illiterate migrant workers are eager to take jobs in a mine, and are powerless to voice safety concerns or establish safety protocols themselves' (Homer 2009: 437). A service worker in a mining community explained the problem: 'The miners are poor people, who have to choose this dangerous job to make a living' (Yue cited in MacLeod 2014);
- conflicts of interest. It is not unknown for public officials to own stakes in mines;

- corruption. According to Homer (2009: 435) 'corruption may involve pay-offs to safety-inspection teams';
- media denialism. According to MacLeod (2014), bad news from home is under-reported while bad news from abroad is over-reported. In August 2014, one of China's deep mines suffered an explosion while another suffered a flood, leaving several dead and thirty-six trapped. Reporting was subdued: 'Although 25 men are trapped underground in Anhui and 11 remain unaccounted for after the mine flooding in Jixi in northeast Heilongjiang province, China's state television broadcaster, CCTV, gave very little coverage to either tragedy. CCTV focuses on positive news that reflects well on Chinese authorities and emphasises negative news about foreign countries. Significant airtime was spent Wednesday on the protests in Ferguson, Mo., over the killing of a young black man by a white policeman' (MacLeod 2014). In totalitarian China, the Fourth Estate's capacity to hold the rich and powerful and corrupt to account is not realised.

Figure 3. Almost 80 per cent of China's coal mines are unregulated. Entrance to a small mine (Wikimedia Commons 2018a).

The actor-network-theory concept *translation* helps explain why so many die in China's deep mines. Miners die because institutions with the potential to hold owners and managers to account have been translated (co-opted) by those who own and manage the mines. Bribery, corruption and conflicts of interest act to stultify oversight. Media jingoism ensures that near-misses, incidents and accidents are either kept off the news agenda or are given tokenistic coverage only. China's mass media has been translated by the State: 'In 2016 [the] France-based watchdog group Reporters Without Borders ranked China 176 out of 180 countries in its 2016 worldwide index of press freedom' (Xu and Albert 2017). Likewise China's trades unions.

Chinese society is un-pluralistic. Those with grievances often have no one to turn to. Actants such as bribery, corruption, conflicts of interest, jingoism and unemployment influence the character, behaviour and safety-performance of China's deep-mining industry. China's deep-mining corruption actor-network, consisting of owners, managers, corrupt politicians, self-interested civil servants, a job-hungry workforce, weak trade unions, censorship, media jingoism, plant, infrastructure, rising demand for electricity and a national culture infected with totalitarianism dominates by translating those considered a threat.

Translation makes the occurrence of latent errors/resident pathogens more likely, because it neutralises safety-assuring checks-and-balances. Ming-Xiao, Tao, Miao-Rong, Bin and Ming-Qiu (2011: 275), in their paper on mines safety, propose the integration of 'scientifically-based safety-protection measures into the production process' as an antidote to near-misses, incidents and accidents. Should such measures threaten the *status quo*, there is every reason to believe that the sponsors of the scientifically based safety-protection measures would be translated, that is, co-opted and neutralised, by China's deep-mining corruption actor-network. In another paper on mines safety, Wei, Hu, Luo and Liang (2017: 80) claim that 'coal mine safety ... can be ... enhanced by ... strengthening supervision'. Such remedies will not work unless the capacity of China's deep-mining corruption actor-network to translate initiatives it considers threatening is neutralised.

Most often, the origins of near-misses, incidents and accidents are complex – embedded in a miasma of social, economic, political and systemic factors. Systems-thinking helps investigators tease-out the decisions and actions that caused a system to fail. Systems-thinking has evolved through numerous iterations. The iteration known as actor-network theory (ANT) conceives of systems as purposive assemblies of ideas and things, each of which possesses agency. Systems-thinking is guided by a simple premise: that to understand the particular one must understand the general. Because of their scope and ambition, investigations informed by systems-thinking are resource-intensive. They can be protracted. They are intellectually demanding. The truth does not come easily, cheaply or quickly.

CHAPTER 2

Systems-thinking in practice

Air accident investigation

Aviation's free-thinking culture has given us many things – a means of defeating the Axis powers, affordable mass-transportation, practical, supersonic air service (for a while) and the systems-thinking-informed approach to incident and accident investigation. Looking back, it is hard to believe that there was a time when air accident investigators either ignored or discounted proximate factors. And it is hard to believe that other high-risk industries (for example, rail transportation, sea transportation, nuclear power generation and offshore oil-and-gas extraction) either ignored or dismissed the systems-thinking-informed approach to investigation. Our disbelief is testament to how far the accident investigation paradigm has shifted.

The 1989 Dryden air disaster

The publication of the Honourable Mr Justice Virgil P. Moshansky's investigation into the 1989 Dryden, Canada, air disaster (Moshansky 1992) was a watershed moment for the systems-thinking-informed approach to accident investigation. In the years that followed, the methodology pioneered by Moshansky and his team permeated the aviation industry and the military (see 'The 2006 Nimrod loss' and 'The 2017 US Navy collisions and grounding' below), then migrated to other sectors such as maritime transportation, rail transportation, nuclear power generation and offshore oil and gas production.

On 10 March 1989, Air Ontario's Captain George Morwood and First Officer Keith Mills were rostered to fly a four-sector day, shuttling between

Winnipeg and Thunder Bay. Morwood had 24,000 flying hours under his belt, and Mills over 10,000 hours. However, neither pilot had much experience on the twin-engined, high-tailed Fokker F28 regional airliner. Morwood had sixty-two hours, and Mills sixty-six hours. The weather was poor. Icing conditions obtained.

The crew had to negotiate a complex flying day: While the final two sectors were uninterrupted, the first two required the aircraft to land at Dryden. The F28's auxiliary power unit (APU) was inoperative (and had been so for five days). An aircraft's APU provides electrical power when it is on the ground. Dryden, the intermediate stop on the first two sectors, had no ground-power units (GPUs). Without a functioning APU and with no GPUs at Dryden, one of the F28's engines would have to remain powered-up when the aircraft was on the ramp. Because Air Ontario's operating procedures proscribed de-icing with an engine running, if Morwood landed at Dryden in icing conditions, he would either have to request a replacement aircraft, improvise a buddy-start or risk a departure *sans* the protection provided by de-icing.

The day's first Winnipeg to Thunder Bay sector, with an intermediate landing at Dryden, went reasonably well. However, the Thunder Bay turnaround was tortuous. Problems included the miscalculation of passenger numbers by Air Ontario Operations, falling behind schedule (not unusual in commercial air service) and passenger complaints. Morwood's decision to offload passengers rather than fuel was overridden by Air Ontario Operations. Defuelling the F28 delayed it further. The mood darkened: 'Evidence from eyewitnesses disclosed that these events changed the good-spirited mood which the flight crew had showed earlier' (Maurino, Reason, Johnston and Lee 1998: 62–63).

By the time the crew began the Thunder Bay to Winnipeg sector, with an intermediate at Dryden, Captain Morwood and his team faced numerous challenges:

- the possibility of icing conditions at Dryden, with associated complications (see above);
- lateness – the F28 departed Thunder Bay about one hour behind schedule;

Systems-thinking in practice

- passengers who were unhappy with the service – some needed to make connecting flights at Winnipeg;
- operating in difficult conditions with limited experience on type – neither pilot had much time on the F28;
- following Air Ontario Operations' overruling of Captain Morwood's decision to offload passengers rather than fuel at Thunder Bay, the crew's realisation that it was not fully in control of the service;
- Captain Morwood's stress level: 'The Dryden report leaves no room for doubt that Captain Morwood was exhibiting distinct symptoms of stress when he landed in Dryden' (International Civil Aviation Organisation 1995: 21);
- the potential for disagreement between Morwood and Mills. Morwood worked for Air Ontario, Mills for Austin Airways. Austin Airways, a northern 'bush' operator, had acquired Air Ontario, a Great Lakes or 'metropolitan' carrier. Given his work-experience, Morwood may have been less comfortable operating in severe weather than Mills. 'The harsher demands of flying in the Canadian north are qualitatively different than those of flying in the south' notes the ICAO (1995: 21).

As the stressors and threats (Reason's (1990, 1997, 2013) latent errors or pathogens) multiplied, the chances of an incident or accident ballooned. At just after mid-day, the F28, by now seventy minutes behind schedule, commenced its take-off roll at Dryden. Failing to gain altitude, the aircraft crashed and caught fire. Twenty-four people died (twenty-one passengers and three crew members). Viewed through the prism of systems-thinking, the disaster had numerous immediate and proximate causes. Immediate causes included:

- an accumulation of snow (that turned to ice) on the forward portion of the F28's wings;
- a contaminated runway;
- a stressed flight-deck;
- the crew's failure to perform a walk-around;
- the crew's decision not to de-ice the F28;

- the Dryden dispatcher's failure to tell the pilots that there was snow on the wings;
- the cabin crew's failure to tell the pilots that there was snow on the wings.

Proximate causes included:

- an operational culture within Air Ontario that was unsuited to jet operations;
- tolerance of open maintenance items (such as the F28's unserviceable APU);
- lack of standardisation;
- training weaknesses and omissions (for example, regarding safety issues associated with wing and runway contamination);
- inadequate CRM training (for example, Moshansky's report noted that the cabin crew failed to challenge the flight-deck on the matter of wing contamination);
- risk-creating rostering (the assembly by Rostering of a flight-deck team with limited type-experience);
- lack of foresight (for example, requiring Captain Morwood to fly an aircraft with an unserviceable APU into an airport with no ground power units in winter);
- cultural friction (the merging of a metropolitan carrier with a bush operator). 'The airline [Air Ontario] ... was the product of the merger of two quite different companies, and two incompatible corporate cultures. ... The effects of such differences were enduring ... and were considered to have had a deleterious effect on crew coordination' note Helmreich and Merritt (2001: xviii);
- employee churn, specifically in regard to managers;
- failure to implement a meritocracy, specifically in regard to managers;
- unfilled posts (creating additional workload);
- overburdened staff;
- weak F28 programme-management;
- Air Canada's tolerance of a double standard in operational practice between it, and its subsidiary Air Ontario. 'The evidence reveals the existence of a double safety standard' claims the ICAO (1995: 24);

- a regulatory authority that lacked the resources to oversee and audit operators to the required standard, and to act on warnings in a timely manner (International Civil Aviation Organisation 1995; Maurino, Reason, Johnston and Lee 1998).

While Captain Morwood must bear responsibility for the acts of commission (such as flying an aircraft with an unserviceable APU into an airport with no ground power units in bad weather) and omission (such as the failure to perform a walk-around) that triggered the disaster, the risk-laden environment in which he found himself on 10 March 1989 was created by Transport Canada's, Air Ontario's, Air Canada's and Dryden's inadequacies (Moshansky 1992; International Civil Aviation Organisation 1995). Dryden was a system accident (see Figure 4).

The 2006 Nimrod loss

On 2 September 2006, a Royal Air Force (RAF) Nimrod (military registration XV230) (see Figure 5) exploded over Afghanistan, killing all on board. Shortly after completing air-to-air refuelling (AAR), the crew received a fire warning (from the bomb bay) and a smoke/hydraulic mist warning (from the elevator bay). Emergency drills were initiated, a MAYDAY message was broadcast and the aircraft was aimed at Kandahar airfield. Engulfed in flames, XV230 exploded at around 3,000 ft.

On 3 September 2006, a Board of Inquiry (BoI) was convened. In addition to identifying the accident's 'probable physical causes' (Haddon-Cave 2009: 28), the BoI identified various contributory factors, including an under-estimation of the risks inherent in airframe modification. The BoI found the Nimrod Safety Case (NSC) wanting. Building on the BoI report, Mr Charles Haddon-Cave QC's (2009) *Independent Review into the Broader Issues Surrounding the Loss of the RAF Nimrod MR2 Aircraft XV230*, identified numerous proximate causes, including:

- significant failings within the system of systems (composed of BAE Systems, QinetiQ and the Nimrod Integrated Project Team) charged with compiling the NSC. Errors of fact went unnoticed or unreported;

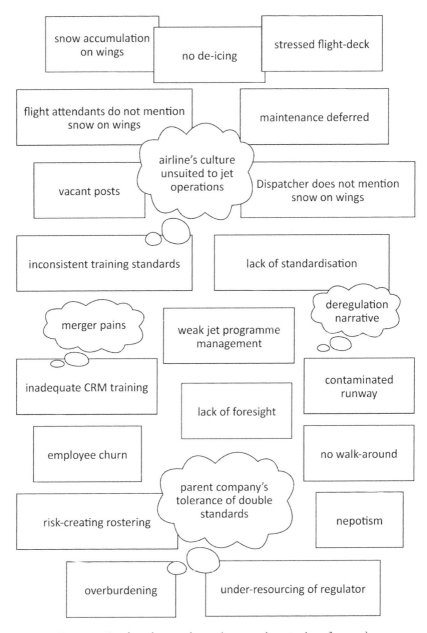

Figure 4. Dryden: the causal soup (not an exhaustive list of actants).

Systems-thinking in practice 25

Figure 5. In happier times: Nimrod XV230 at the 2005 Waddington Air Show, England (Wikimedia Commons 2018b).

- a failure to learn from past incidents, such as the November 2004 rupture of the Supplementary Cooling Pack (SCP) duct in Nimrod XV227 (the fire that brought down Nimrod XV230 in 2006 is thought to have started when an element of the SCP ignited spilled fuel);
- a failure to respond to safety warnings. In 1998, the Nimrod Airworthiness Review Team (NART) highlighted the problem of 'ever-reducing resources and ... increasing demands; whether they be operational, financial, legislative or merely those symptomatic of keeping an old aircraft flying'. The NART warned of an 'incipient threat to safe standards' (the NART cited in Haddon-Cave 2009: 356). Nimrod XV230 was the first Nimrod to enter service with the Royal Air Force, being accepted on 2 October 1969;
- government-initiated reform of the Ministry of Defence's (MoD's) procurement and asset-management practices that, according to Haddon-Cave (2009: 355), destabilised the MoD: 'The MOD suffered a sustained

period of deep organisational trauma between 1998 and 2006 due to the imposition of unending cuts and change, which led to a dilution of its safety and airworthiness regime and culture, and distraction from airworthiness as the top priority';
- the emergence of an organisational culture that prioritised financial efficiency over safety. According to Haddon-Cave (2009: 445), the MoD's new culture 'allowed "business" to eclipse Airworthiness'.

The circumstances that contributed to the loss of XV230 gestated over decades. Contributing to the loss were decisions, acts of commission and acts of omission. For example:

- the decision by Harold Wilson's Labour government not to design a jet-powered anti-submarine aircraft from scratch, but to adapt an existing airframe (that of the de Havilland Comet airliner, designed in the 1940s). France's Breguet had a bespoke design available for purchase (the twin-turboprop Atlantic);
- the installation in the late 1970s of SCPs that, according to Haddon-Cave (2009: 15) 'increased the potential for ignition';
- the addition in the early 1980s of an air-to-air refuelling capability (in preparation for the Falklands War) that, according to Haddon-Cave (2009: 15) 'increased the risk of an uncontained escape of fuel';
- the defence establishment's failure to address the potential safety impacts of reactive patching (see Weir (1996) for a definition of reactive patching);
- the defence establishment's failure to address the issue of the Nimrod aircraft's safety migration (see Rasmussen (1997) for a definition of safety migration);
- the apparent failure of the service's error-reporting systems to trigger effective remediations. It is almost inconceivable that senior officers were unaware of the Nimrod aircraft's spilled-fuel/SCP latent error/ resident pathogen.

Large, complex socio-technical systems are prone to inconsistent performance, upset and breakdown (Shorrock, Leonhardt, Licu and Peters 2014).

Britain's defence establishment, consisting of the armed services, MoD, research laboratories, research agencies, defence equipment suppliers, trade bodies and other elements, constitutes a large, complex socio-technical system-of-systems. Component elements of the system-of-systems may pursue divergent agendas, creating incoherence, conflict and risk. Pressured work-teams may develop a committed world-view that excludes alternative interpretations of facts. Work-teams may go so far as to ostracise, or attempt to discredit, those who offer alternative interpretations (a dynamic identified by Janis (1972) as groupthink).

To a degree, incidents and accidents are a product of the times in which they occur. The loss of XV230 occurred in a context of budget cuts, organisational perturbation and military challenge. During its time in service, the Hawker Siddeley/British Aerospace/BAE Systems Nimrod fleet had to meet a variety of operational demands, including:

- preparing for a hot war with the Soviets. In time of war 'No 18 (Maritime) Group [would be responsible for] reconnaissance; support of the Atlantic Striking Fleet; control, routing and protection of shipping; and offensive operations against enemy submarines' (Fricker 1975: 9);
- liberating the Falklands;
- supporting NATO in the Balkans;
- participating in Operation Herrick;
- participating in Operation Iraqi Freedom.

To meet the various demands placed on the Nimrod design, the MoD provided funds to Hawker Siddeley/British Aerospace/BAE Systems for upgrades. For example, the upgrade to MR2 standard included: '[N]ew EMI radar, a new sonics system, a navigation system of improved accuracy, increased computer capacity and improved display system techniques' (Green 1975: 104). Unfortunately, over time, the upgrades, or reactive patches, degraded system integrity. The 2 September 2006 fire foregrounded the airframe's loss of integrity. Incidents, accidents and near-misses are *revelatory*. Often they foreground vulnerabilities that have existed, unseen and unappreciated, for years (see Figure 6).

Figure 6. RAF Nimrod XV230 loss: the causal soup (not an exhaustive list of actants).

Aviation has pioneered the systems-thinking approach to incident and accident investigation. There have been bumps in the road, such as the authorities' attempt to limit the terms of reference of Mr Justice Virgil P. Moshansky's investigation into Dryden. The effort to undermine Moshansky's investigation spoke to the power of the methodology. Those with something to hide feared it. The Moshansky and Haddon-Cave investigations proved that disasters gestate. The Dryden accident originated in systemic weaknesses. The Nimrod loss originated in decades of airframe modification necessitated by new challenges. In each case, politics helped sow the seeds. In the case of Dryden it was economic liberalisation. In the case of the Nimrod loss it was geopolitics. By expanding the problem-space, systems-thinking-informed investigations take in less obvious, but no less important causative or contributory factors such as political objectives, shareholder agendas and geopolitics.

Marine accident investigation

Over the years, the systems-thinking-informed approach to incident and accident investigation has been adopted by several industries, including marine transportation. The systems-thinking-informed approach is enshrined in the standards and recommended practices (SARPs) of the United Nations' International Maritime Organisation (IMO) (2008: 17): 'Proper identification of causal factors requires timely and methodical investigation, going far beyond the immediate evidence and looking for underlying conditions, which may be remote from the site of the marine casualty or marine incident, and which may cause other future marine casualties and marine incidents. Marine safety investigations should therefore be seen as a means of identifying not only immediate causal factors but also failures that may be present in the whole chain of responsibility'.

The IMO's systems-thinking-informed approach to incident and accident investigation informs how investigations are conducted by memberstates (see Figure 7). For example, the American Bureau of Shipping's (ABS's) (2005) *Guidance Notes on the Investigation of Marine Incidents*, recommends the investigation of both immediate and proximate causes:

[During an investigation], human errors, problems (including those related to structure, machinery, equipment or outfitting items) and/or external factors are analysed. Farther down in the triangle are more fundamental causes and aspects of organisations. These include controls for the task and for the process. Eventually, management systems and the organisation's culture can be analysed. Analysing deeper into the triangle allows organisations to increase the level of learning about how the organisation functions and, therefore, develop corrective and preventative actions that are more fundamental in nature and broader in scope. *These fundamental changes allow problems to be solved once instead of several times.* (American Bureau of Shipping: 12, my emphasis; see Figure 7)

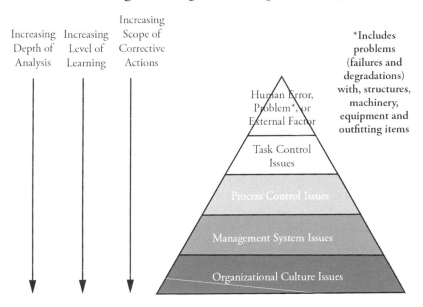

Figure 7. Possible depths of analyses in a systems-thinking-informed investigation (American Bureau of Shipping 2005: 13).

As explained in Figure 7, the more comprehensive one's investigation of proximate causes, the greater the level of learning and potential for corrective actions to prevent a repeat. In the United States, both commercial and non-commercial (naval) marine incidents and accidents are subject to systems-thinking-informed investigations.

The 2017 US Navy collisions and grounding

The year 2017 was an eventful one for the United States Navy. On 31 January 2017, the USS *Antietam*, a 10,000 ton guided-missile cruiser, ran aground off the coast of Japan, damaging its propellers. On 9 May 2017, a South Korean fishing smack collided with the USS *Lake Champlain*, a 10,000 ton guided-missile cruiser. On 17 June 2017, the USS *Fitzgerald*, a 9,000 ton guided-missile destroyer, collided with a 730ft-long, 40,000 ton container ship. The *Fitzgerald* sustained significant damage. Seven US Navy sailors were killed and three injured (including the Captain, whose cabin was crushed in the collision). On the morning of 21 August 2017, the USS *John S. McCain*, a 9,000 ton guided-missile destroyer (see Figure 8), collided with the *Alnic MC*, a 600ft-long, 30,000 ton oil-and-chemical tanker, off the coast of Singapore. The *McCain* sustained significant damage (see Figure 9), water flooding into various compartments, including machinery and communications rooms and crew berthing. Ten US Navy sailors were killed and five injured (Department of the Navy 2017).

Figure 8. The USS *John S. McCain* (Wikimedia Commons 2018c).

While United States Navy warships have been involved in numerous, deadly, at-sea collisions (for example, the 1975 collision between the 9,000 ton guided-missile cruiser USS *Belknap* and the 83,000 ton aircraft carrier USS *John F. Kennedy* that killed eight sailors and saw the *Belknap* burn for several hours, and the 2001 collision between a Los Angeles-class attack submarine and a Japanese fishing boat that killed nine, including four high-school students), the concentration of incidents and accidents in 2017 prompted the Department of the Navy to order a thorough investigation of the service's operational readiness. The investigation team included experts from industry and academia. Taking sixty days, the investigation was tasked 'to identify gaps in Doctrine, Organisation, Training, Material, Leadership and Education, Personnel and Facilities (DOTMLPF)' (Department of the Navy 2017: 6–7). The resulting *Comprehensive Review of Recent Surface Force Incidents* supplemented the investigations conducted into individual mishaps.

Drawing on systems-thinking, the *Comprehensive Review* identified a range of immediate and proximate causes. These included:

- poor seamanship and departures from safe practice: 'In each case [personnel] did not maintain situational awareness. … In every mishap [there were] departures from procedures or approved customary practices' (Department of the Navy 2017: 16);
- poor teamwork;
- poor team-management: 'Command leadership … failed in their absolute responsibility to … properly balance their watchteams with depth and experience to foster consistent superior performance' (Department of the Navy 2017: 16);
- insufficient time available for training due to '[t]he increase in operating tempo in the Western Pacific combined with longer … maintenance and modernisation periods' (Department of the Navy 2017: 17);
- lax oversight by headquarters staff;
- inadequate critical assessment (that is, inadequate self-critique or reflexivity);
- inadequate safety programmes and flawed systems for safety-reporting and analysis;
- inadequate trend-analysis;

- management decisions informed by erroneous assumptions, for example, 'that a high pace of operations equates to a high state of proficiency' (Department of the Navy 2017: 19);
- decision-making informed by a can-do navy culture that induced 'a slow erosion of standards, and organisational drift from the deliberate processes used to manage time, resources [and] rest' (Department of the Navy 2017: 19). Crew fatigue posed a threat to safety;
- under-funded modernisation programmes that created 'wide variations in [sensor and control] configurations from ship-to-ship' (Department of the Navy 2017: 19). Variation of configuration creates latent errors/resident pathogens – traps for the unwary;
- risk-blindness and poor-quality decision-making at the highest levels of command: 'Even when presented with information that indicated standards of readiness were not met, rather than taking pause ... the force was conditioned to mitigate the risk ... through the delay of some training action that would ... lessen the impact on operational missions' (Department of the Navy 2017: 20). Put another way, when presented with evidence that things were not as they should be, commanders chose the least effective remediations.

The 2017 mishaps were, in varying degrees, system accidents. While those who sailed the vessels made mistakes, those whose responsibility it was to mitigate operational risk (by, for example, training ratings to the required standard, maintaining skills, creating experientially balanced work-teams, ensuring that sailors were not physically or psychologically over-extended and ensuring the timely analysis of failure-trends) failed to discharge that responsibility adequately. As the saying goes, safety is everyone's responsibility.

> The Department of the Navy's *Comprehensive Review of Recent Surface Force Incidents* was a model investigation. Instead of blaming crewmen, the *Comprehensive Review* presented a balanced analysis of proximate and contextual factors. While accepting that errors were made, it foregrounded systemic failings such as underfunding, lack of standardisation and high-tempo operations that disrupted training and left crews jaded and fatigued. It would have been easy for the authors of

the *Comprehensive Review* to scapegoat officers and ratings. Instead, the authors of the *Comprehensive Review* chose the more difficult and effortful route of identifying latent errors/resident pathogens in the *modus operandi* of the Navy and Department of Defence. Conducting the investigation in this way required a determination to get at the truth. Deep-analysis is effortful and not without risk. When subject to investigation, the powerful may take offence. They may lash out. Power corrupts.

Figure 9. A severely damaged USS *John S. McCain* limping towards Changi Naval Base in the Republic of Singapore (Wikimedia Commons 2018d).

Rail accident investigation

The Lac-Mégantic derailment and fire

The Montreal, Maine and Atlantic Railway (MMA) train involved in the derailment and fire, made up of seventy-two tank cars and five locomotives, was single-crewed. Late on 5 July 2013 the engineer parked the train on an incline at Nantes, Quebec. Having applied seven sets of hand-brakes (instead of the required nine), he tested the configuration with the train's air-brakes powered-up. The engineer should have tested the configuration with the air-brakes powered-down: 'That night … the locomotive air brakes were left on during the test, meaning the train was being held by a combination of hand brakes and air brakes. This gave the false impression that the hand brakes alone would hold the train' (Transportation Safety Board of Canada 2014a: 2). After setting the brakes, the engineer reported that the lead locomotive had developed a fault (black and white smoke had been issuing from the engine exhaust). He retired for the night.

During the night, the lead locomotive caught fire. The rail company sent a track foreman to support the Nantes Fire Department in its efforts to save the locomotive. The fire was extinguished and the locomotive powered-down. As air leaked from the train's air-brake system, the unattended train began to travel in the direction of Lac-Mégantic and its 6,000 inhabitants, reaching a top speed of 65 mph. It derailed in the town at about 01:15, the ensuing fire and explosions causing significant loss of life and destruction of property: 'Almost all of the 63 derailed tank cars were damaged, and many had large breaches. About six million litres of petroleum crude oil was quickly released. The fire began almost immediately, and the ensuing blaze and explosions left 47 people dead. Another 2,000 people were forced from their homes, and much of the downtown core was destroyed' (Transportation Safety Board of Canada 2014a: 3). The Fire Department performed heroically. Had the derailment occurred during the day, the death toll might have been higher. Applying the Challenger, Clegg and Robinson (2009) model of system function, causal factors of the Lac-Mégantic rail disaster included:

GOALS

Like all enterprises, railway companies seek to reduce costs. To this end, maintenance items may be deferred, and repairs, when done, may be gerrymandered. According to the Transportation Safety Board of Canada (TSBC) (2014a: 4), the fire in the lead locomotive resulted from a rushed, gerrymandered repair: 'In October 2012, eight months before this accident, the lead locomotive was sent to MMA's repair shop following an engine failure. Given the *significant time and cost of a standard repair, and the pressure to return the locomotive to service*, the engine was repaired with an epoxy-like material that lacked the required strength and durability. This material failed in service, leading to engine surges and excessive black and white smoke. Eventually, oil began to accumulate in the body of the turbocharger, where it overheated and caught fire on the night of the accident [my emphasis]'. It was this fire that triggered the events that led to the MMA train derailing and igniting in downtown Lac-Mégantic.

In its report into the disaster, the TSBC criticised both the quality of the rail company's oversight and the quality of its training programmes: 'There were … significant gaps between the company's operating instructions and how work was done day to day. … [The TSBC] found that employee training, testing and supervision were not sufficient, particularly when it came to the operation of hand-brakes and the securement of trains'. Possibly, the MMA's determination to cut costs had compromised oversight and training.

The TSBC questioned the MMA's route-planning, referring to 'the risk of not planning and analysing routes on which dangerous goods are carried' (Transportation Safety Board of Canada 2014a: 11). The MMA would have been aware that direct routing minimises cost.

Finally, the TSBC noted that the train was made up of tank-cars built to an earlier, less crash-worthy standard (Transportation Safety Board of Canada 2014a). The MMA's determination to cut costs may have influenced it's decision not to improve the cars.

TECHNOLOGY

The design and maintenance of the equipment left much to be desired:

- the tank-cars were built to an older, less crash-worthy design standard. According to Campbell (2013b: 6–8), '[T]he rail industry ha[d] met [a] spectacular surge in demand, for the most part, using tank cars that were not built for carrying hazardous materials. … Regulators … have known since the early 1990s that the … tanker cars had a propensity to

puncture during derailments. The DOT-111 is an all-purpose tanker car with a single steel shell, and was not designed to carry hazardous products. The Transportation Safety Board documented its concerns repeatedly';
- the repair to the lead locomotive had been gerrymandered, creating a latent error or resident pathogen (a potential failure).

Despite these failings, the chairman of the Montreal, Maine and Atlantic Railway Ltd insisted that the disaster was entirely the fault of the engineer. Said Edward Arnold Burkhardt: '[The residents of Lac-Mégantic] view me terribly, but I wasn't the guy who didn't set the brakes on the train' (Burkhardt cited in Canadian Broadcasting Corporation News 2013). As far as the chairman was concerned, there were no corporate failings: 'There's always going to be risks in handling dangerous commodities. You try to minimize those risks, you try to manage your company well, so that those risks are low' (Burkhardt cited in Canadian Broadcasting Corporation News 2013). The reductionist explanation offered by Edward Burkhardt was challenged by the chairperson of the TSBC, Wendy Tadros (cited in Tadros 2014), who told a news conference: '[T]he chain of causes and contributing factors goes far beyond the actions of any single person'.

BUILDINGS AND INFRASTRUCTURE

Because the track had been laid through the heart of downtown Lac-Mégantic, any mishap would always have the potential to cause loss of life and/or collateral damage. In the event, the fires and explosions killed forty-seven and obliterated approximately half of Lac-Mégantic's downtown area. Chronic impacts included benzene contamination of the soil and waterways. Many businesses relocated. The authorities drew up plans to replace contaminated soil with clean-fill. Running a railway track through a neighbourhood creates risk. Routing flammable cargoes along that track adds risk. To a certain extent, the Lac-Mégantic disaster was constructed (created). Decisions have consequences.

In May 2018, the government approved a $133-million (£79-million) project to build a rail bypass for Lac-Mégantic: 'The federal government will foot 60 per cent of the bill for the construction of the 11-kilometre [7-mile] track, with Québec covering the remainder of the cost. ... Construction work will get underway in 2019 and is expected to take two years to complete' (Brunette 2018).

PROCESSES AND PROCEDURES

There were questionable practices. At the MMA these included:

- the gerrymandering of repairs (discussed above). In its final report, the TSBC criticised '[The] MMA management's tolerance of non-standard repairs ... which either subsequently failed, or did not return the parts to their proper operating condition' (Transportation Safety Board of Canada 2014b: 124);
- the rostering of a single engineer to a five-engine, 72-car train carrying flammable payload, specifically 7.7 million litres (1.7 million gallons) of petroleum crude oil. An exemption allowed the MMA to operate its freight trains in this manner: 'The Transport Minister granted MMA an exemption from the required two-person crews – one of only two exemptions granted for a freight railway – despite objections from the union ... and its troubling safety record – possibly due in part to pressure to adopt the lower US standard – which permits one-person crews' (Campbell 2013a);
- the routing of flammable payload through urban areas (discussed above);
- leaving trains carrying flammable payload unattended overnight;
- parking such trains on a gradient.

Questionable practices at Transport Canada included:

- the arms-length supervision of rail operators such as the MMA. Regarding the performance of Transport Canada, the TSBC (2014a: 10) noted: 'Inadequate oversight of operational changes; Limited follow-up on safety deficiencies; Ineffective SMS audit programme'. According to Campbell (2013a), the regulatory regime in play at the time of the Lac-Mégantic disaster allowed 'profit-seeking companies to largely regulate themselves'. Campbell (2013b: 7) observes: 'Over the last 25 years, Transport Canada has increasingly devolved the responsibility

Systems-thinking in practice

for, and management of, safety rules to the companies themselves. ... Amendments to the Railway Safety Act ... enacted in 2001 gave companies the authority to implement safety management systems ... enabling companies to develop their own rules and standards. They allow companies to make their own judgements about the balance between cost considerations and the risks to public safety. Under this system, federal inspectors audit and approve the SMS ... but carry out far fewer on-site inspections. Referred to as co-regulation between government and industry, this is, in effect, self-regulation – a major surrender of Transport Canada's regulatory authority';
- tolerance of reactive risk-management at the MMA. The most risk-aware companies manage risk proactively: 'An organisation with a strong safety culture is generally proactive when it comes to addressing safety issues. MMA was generally reactive' (Transportation Safety Board of Canada 2014a: 10);
- tolerance of recurring safety issues at the MMA. Recurrence suggests a failure by the regulator and the MMA to address latent errors/resident pathogens (such as the company's weak safety culture);
- failure of Transport Canada's Quebec regional office to audit the MMA's safety management system (SMS) in a timely manner. Despite being drafted in 2002, the MMA's SMS was not audited until 2010, at which point it was implemented (Transportation Safety Board of Canada 2014b);
- failure of Transport Canada to evaluate rail companies' safety culture: 'Although educational material about safety culture was provided to railway companies, safety culture was not formally assessed ... within regulatory inspections' (Transportation Safety Board of Canada 2014b: 124);
- failure of Transport Canada's Headquarters in Ottawa to effectively monitor and, where required, improve the oversight performance of its Quebec regional office (Transportation Safety Board of Canada 2014a).

There were various procedural failures. For example:

- the shipping documents for the MMA train misrepresented the volatility of the cargo: 'The petroleum crude oil in the tank cars was more volatile than described. ... If petroleum crude oil is not tested systematically and

frequently, there is a risk of it being improperly classified. The movement of these improperly classified goods increases the risk' (Transportation Safety Board of Canada 2014a: 8);
- the *Canadian Rail Operating Rules* required that nine hand-brakes be set for a seventy-two-car train. Only seven were set by the engineer at Nantes – a violation. There was no second engineer to cross-check the actions of the first. Cross-checking is one of the pillars of aviation safety.

Such practices and failures suggest a lax safety culture at the MMA. As noted by the TSBC (2014b: 124): 'If instructions or rules are disregarded, and unsafe conditions and practices are allowed to persist, this leads to an increased acceptance of such situations. Deviations from the norm thus become the norm, and the likelihood of unsafe practices being reported and addressed is reduced. … MMA's weak safety culture contributed to the continuation of unsafe conditions and practices, and compromised MMA's ability to effectively manage safety'.

CULTURE

Culture is 'the way we do things *here*' (Helmreich and Merritt 2001: 1). From a safety perspective, the way Transport Canada and the MMA conducted themselves left much to be desired. The organisational culture of each was dysfunctional. Failings at Transport Canada compounded those at the MMA. The Montreal, Maine and Atlantic Railway Ltd's failings went undetected or uncorrected. In hindsight, Transport Canada and the MMA were locked in a downward spiral of lax oversight, cost-cutting and dangerous practices. With reference to Reason's (1990, 1997, 2013) Swiss cheese model of failure, the circumstances of 5/6 July 2013 breached the MMA's weak defences, provoking disaster.

At the time of the disaster, rail-freight was booming: 'Shipments of oil and gas by rail tank cars grew by 14.3 percent in 2013. … This increase is largely attributed to rapid increases in crude production in both Canada and the United States, stagnant current North American pipeline capacity and revenue opportunities in diversified markets … The number of crude oil rail carload movements has increased significantly over the last couple of years, from around 340 in 2010 to more than 53,000 in 2012' (Transport Canada 2014: 18).

Despite the boom, the MMA evolved a culture of cost-cutting. Campbell (2013b: 7) notes: 'Changes to the Canada Transportation Act in the 1990s spurred a major restructuring of large carriers, resulting in a proliferation of smaller railways. One such line, the Montreal, Maine and Atlantic ... embarked on a drastic cost-cutting exercise, laying off staff and cutting wages, in an effort to turn a profit'. Wendy Tadros, TSBC chairperson, described the MMA as 'A short-line railway running its operations at the margins. Choosing to lower the track speeds rather than invest more in infrastructure. Cutting corners on engine maintenance and training' (Tadros 2014).

While management failings within Transport Canada (TC) and the Montreal, Maine and Atlantic Railway helped create the conditions for disaster, the broader political narrative also contributed. For example:

- governments' determination to exploit oil reserves fomented a 'wild-west boom in the production of unconventional oil' (Campbell 2013b: 6). This created a logistical problem for both the United States and Canada: specifically, how to transport large volumes of crude oil over large distances *in absentia* a pipeline network? Despite being ill-equipped for the task, rail provided a fall-back. In the early years of the boom, all went well: 'Over the last decade, accidents where dangerous goods spilled have been few, and have remained stable or even declined slightly' (Campbell 2013b: 10). Because little had gone wrong during the early years of the boom, it is likely that actors such as Transport Canada, the Railway Association of Canada and the rail companies assumed that nothing *could* go wrong. It is likely that actors deduced from a generally satisfactory safety record that companies' safety management systems were performing as required, that is, that the SMSs were ensuring that operators struck a socially responsible balance between safety and efficiency. The Lac-Mégantic disaster suggests that actors' confidence in the industry's *modus operandi*, specifically, in the efficacy of self-regulation, was misplaced;
- the global penchant for arms-length regulation persuaded legislators that it was possible to manage safety through the medium of company-specific safety management systems. According to Campbell (2013b: 7),

TC's enthusiastic promotion of SMSs served to undermine its authority: 'Amendments to the Railway Safety Act ... gave companies the authority to implement safety management systems ... enabling companies to develop their own rules and standards. They allow companies to make their own judgements about the balance between cost considerations and the risks to public safety. Under this system, federal inspectors audit and approve the SMS ... but carry out far fewer on-site inspections. Referred to as co-regulation between government and industry, this is, in effect, self-regulation – a major surrender of Transport Canada's regulatory authority';

- the Canadian government's determination to reduce the deficit saw TC's budget slashed and its capabilities reduced: 'In its austerity drive ... the Harper government has not spared rail safety. Conservative budgets from 2010–2011 to 2013–2014 – a period of enormous expansion in oil-by-rail traffic – slashed the rail safety budget by 19%. Transport Canada also shaved the very small Transportation of Dangerous Goods budget over those four years ... Moreover, in spite of expectations that oil transport by rail will continue to grow rapidly, it plans to freeze the budgets for both divisions thereafter until at least 2015–2016' (Campbell 2013b: 7). The austerity programme bore down on TC's capacity to oversee rail-freight: 'The Conservative government did not increase the number of inspectors to handle the enormous increase in oil-by-rail traffic. ... While in 2009 there was one inspector for every 14 tank carloads of crude oil, by 2013 there was only one inspector for every 4,000 tank carloads' (Campbell 2013b: 9).

PEOPLE

The chairman of the MMA laid the blame entirely on the shoulders of his engineer: 'If you have a situation where the engineer violates the rules ... it creates ... a tragic situation. The fact is this is a failure of one individual' (Burkhardt cited in Mackrael and Robertson 2014). Burkhardt's reductionist analysis was rejected by the TSBC and, unsurprisingly, by the lawyer acting for the engineer who left the train unattended: 'I think what it means for [MMA engineer Tom Harding] and the other people that are accused ... is that the public is aware now that their role in that thing was a partial role' (Walsh cited in Mackrael and Robertson 2014). Tom Harding erred. However, he did so in a context of aggressive cost-cutting, weak supervision, sub-standard training, stigmatisation of safety rules, arms-length regulation, a political desire to exploit a once-in-a-generation energy opportunity (Campbell 2013b) and the public's love affair with the automobile.

CONCLUSIONS

In the aftermath of the Lac-Mégantic rail disaster, the MMA's chairman rounded on engineer Harding, ignoring mounting evidence that Lac-Mégantic was a system accident, rooted in ineptitude, indifference, corporate greed and national ambition. For a chairman in denial and a shocked nation seeking answers, Harding was a convenient scapegoat.

The seeds of Lac-Mégantic were sown years before the night Harding walked away from his train. Those seeds included aggressive lobbying, political capitulation, deregulation, gullibility, disempowerment of the state and tolerance of bad practice in pursuit of a larger good – the revitalisation of the Canadian economy.

Engineer Tom Harding suffered the same fate as pilot James Thain. Like Thain, Harding was victimised for his proximity to a system accident.

> Every incident, accident and near-miss has a trajectory consisting of historic decisions, actions and episodes. It is not possible to understand why Tom Harding walked away from his malfunctioning locomotive without understanding Canada's philosophical direction of travel and the culture and practices of its rail industry. Vested interests induce actors to attribute disasters to different causes. The MMA's chairman, seeking to protect his company's reputation and his own name, blamed engineer Harding. In 1958, the British government, seeking to rebuild relations with Germany, blamed Captain Thain. In the aftermath of the 1989 Hillsborough football stadium disaster that killed ninety-six fans, South Yorkshire Constabulary altered statements to cover-up police failings. History is peppered with examples of the weak and vulnerable being sacrificed by the rich and powerful to promote an interest. As The Right Reverend James Jones KBE observed in his report into the Hillsborough cover-up (Jones 2017: 6), individuals and institutions often act in their own rather than the public

interest: '[W]hat I describe ... as "the patronising disposition of unaccountable power" ... does not just describe the families' experience of the police, but also of other agencies and individuals across the criminal justice system and beyond. And it does not simply describe a historic state of affairs, but instead one that stretches forward to today ... What this report describes as a "patronising disposition" is a cultural condition, a mindset which defines how organisations and people within them behave and which can act as an unwritten, even unspoken, connection between individuals in organisations. *One of its core features is an instinctive prioritisation of the reputation of an organisation over the citizen's right to expect people to be held to account for their actions* [my emphasis]'.

Nuclear accident investigation

The 2011 Fukushima Daiichi nuclear accident

The 11 March 2011 Tōhoku undersea earthquake (epicentre: approximately 43 miles (69 kilometres) east of Japan's Oshika Peninsula of Tōhoku; hypocentre: approximately 18 miles (29 kilometres) under the ocean) set in motion a series of events that disabled the six-reactor Tokyo Electric Power Company (TEPCO) Fukushima Daiichi oceanside nuclear power plant. The earthquake and tsunami caused power outages, chemical explosions, meltdowns, damage to buildings and radioactive emissions at the TEPCO site. Land was flooded and agricultural output lost (see Figure 10).

Fukushima was the worst nuclear accident since Chernobyl. Impacts included the contamination of 6,950 square miles (1,800 square kilometres) of land and the evacuation of 150,000 people. Decontamination, decommissioning and disposal will take several decades. As shown by Britain's efforts to dismantle the Dounreay experimental fast-breeder reactor, decommissioning a nuclear facility is a complex, costly and risk-laden process. Decommissioning a nuclear facility that has been ripped apart presents

Systems-thinking in practice 45

an even greater challenge. The Chernobyl decommissioning testifies to the difficulties the Japanese can expect. In 2018, Ukraine announced the commissioning of Chernobyl's New Safe Confinement (NSC) shelter. The European Bank for Reconstruction-funded edifice cost a staggering €1.5bn ($1.7bn). The NSC's 100-year lifespan will 'allow for the eventual dismantling of the ageing makeshift shelter from 1986, and the management of the radioactive waste' (Dalton 2018a).

Figure 10. Helicopter mission over the inundation caused by the tsunami (Wikimedia Commons 2018e).

Following the disaster, The National Diet of Japan (2012), to its credit, conducted a systems-thinking-informed investigation into events before, during and after the 11 March sub-sea earthquake, tsunami, inundation, reactor meltdowns, chemical explosions and releases. Analyses produced by other bodies, such as the International Atomic Energy Agency (IAEA) (2015) and World Nuclear Association (WNA) (2016) also drew on systems-thinking. A consensus formed around the immediate and proximate causes of the disaster. Applying the Challenger, Clegg and Robinson (2009) model of system function, causal factors of the Fukushima Daiichi nuclear accident included:

GOALS

Systems theory recognises that environmental factors impact behaviour at multiple levels, including at the level of the nation-state. As Monteiro (2012) puts it: 'You do not go about doing your business in a total vacuum, but rather under the influence of a wide range of … factors'. Historically, Japan's lack of essential raw materials and consequent reliance on imports has shaped its foreign and domestic policy. Japan's twentieth-century foreign adventures, such as the invasions of Manchuria, China and French Indochina, were spurred by her desire to secure reliable supplies of raw materials (Ambrose 1985). The economic depression of the 1930s stirred Japan's militarists and steeled the will of the people (Columbia University 2009). The country was prepared to do what it considered necessary for survival.

In the twenty-first century, the Japanese government has used soft power – loans, grants and offers of technical co-operation – to woo oil and gas-rich nations (Pollmann 2016). Japanese Prime Minister Shinzo Abe's 2016 state visit to gas-rich Russia spoke to Japan's desire to expand its pool of energy suppliers. Japan's strategic energy mix includes oil, natural gas, clean coal, wind power, solar power and nuclear energy. Japan's lack of essential raw materials and consequent dependence on energy imports shapes its world-view. To put it another way, Japan looks at the world through a poverty-of-raw-materials prism.

For the Japanese, energy security is synonymous with national security. Framed as a cornerstone of energy security by successive governments, nuclear power has been enthusiastically promoted. The government's nuclear policies reflect the country's historic and current depredations and vulnerabilities. In relation to energy policy, Japan is a prisoner of circumstance.

TECHNOLOGY

Although nuclear accidents are rare, when they do occur, as at Windscale in 1957 (see Figure 11), Three Mile Island in 1979 and Chernobyl in 1986, the

Systems-thinking in practice 47

Figure 11. Windscale – site of Britain's most serious nuclear accident (Wikimedia Commons 2018f).

consequences can be serious and enduring. Impacts include worker fatalities, an elevated cancer risk, radioactive contamination of the food chain and reputational damage. In 1947, Britain's Labour government announced the construction of an atomic energy complex at Sellafield in remote Cumbria. Sellafield had previously hosted an ordnance factory.

In 1956, the world's first commercial nuclear power station was commissioned on the site (which was renamed Windscale). The power station produced electricity for the National Grid and plutonium for Britain's nascent atomic weapons programme. On 10 October 1957, a Windscale reactor overheated and caught fire. Radioactive iodine 131 was released into the atmosphere. Harold Macmillan's Conservative government implemented short and long-term remediations. Short-term remediations included a 13 October ban on the sale of milk from local farms. Long-term remediations included the passing of the 1959 Nuclear Installations Act, that 'required that civil nuclear power stations which were then under construction, and those planned for the future, be licensed by the newly formed Nuclear Installations Inspectorate ... a regulator whose sole responsibility would be safety' (House of Commons Science and Technology Committee 2012: 7). A case of locking the stable door after the horse had bolted?

A Committee of Inquiry (CoI) concluded there had been 'no district radiation or inhalation hazard' and that the Windscale fire had 'no bearing on the safety of nuclear power stations being built for electricity authorities' (CoI cited in British Broadcasting Corporation 2018b). The 1957 Windscale fire 'remains the most severe nuclear accident in UK history' (House of Commons Science and Technology Committee 2012: 7).

There is disagreement as to how production systems can be made to operate reliably. Some argue that simplicity delivers reliability (McIntyre 2000); others that sophistication (at both the technical and organisational levels) delivers reliability (Roberts 1990; LaPorte and Consolini 1991). Designers of nuclear plant are devotees of the latter approach. They look to system redundancy (defence-in-depth or the belt-and-braces approach) to create safety margins.

The application of this philosophy at Fukushima saw the provision of two types of on-site back-up for the reactors' cooling systems – diesel generators and DC batteries. Unfortunately, most of the plant's back-up systems (generators, batteries and switchgear) were located in buildings that were vulnerable to flooding, creating a latent error or resident pathogen.

At the time of the disaster, the disposition of Fukushima's back-up power systems was as follows:

- Reactors 1–5: Two diesel generators and banks of DC batteries for each reactor. The switching stations were located in the turbine buildings;
- Reactor 6: Three diesel generators and banks of DC batteries located in the turbine building. The switching station for this reactor's back-up electrical energy system was located in the watertight reactor building (this was untypical for the site);

- Reactors 1–6: During the late 1990s, three additional diesel generators had been installed in a less-vulnerable hillside building. However, the switching stations for reactors 1–5 remained in the turbine buildings.

At the time of the earthquake and tsunami, only reactors 1, 2 and 3 were operating. Reactors 4, 5 and 6 were undergoing inspection and maintenance. The earthquake and tsunami, which reached 46–49 feet (14–15 metres), caused extensive damage:

- Fukushima's off-site power connectors went down;
- twelve generators were disabled;
- the switchgear was disabled;
- three reactors suffered melt-downs due to coolant loss;
- overheating caused explosions which released radioactivity into the environment (The National Diet of Japan 2012; House of Commons Science and Technology Committee 2012; International Atomic Energy Agency 2015; World Nuclear Association 2016).

The tsunami rendered Fukushima's latent errors active, disabling the plant's back-up power supply. In complex socio-technical systems such as aircraft, naval vessels or nuclear power plants, the greater the number of design weaknesses, the greater the likelihood of a system accident.

For their potential to be realised, back-up systems must be protected against foreseeable adverse events. The installation of three additional generators in the late 1990s (to meet a new safety standard) created a window of opportunity to improve the resilience of the plant's back-up electrical energy system (that is, there was an opportunity to harden the back-up system). This opportunity was not taken.

Finally, as mentioned above, nuclear power generation creates unique risks, the manifestations of which (irradiation, environmental contamination, resource-intensive clean-ups, risks associated with the long-term storage of waste, reputational damage, negative perceptions of the nuclear industry, etc.) *endure*. An accident at a conventional gas, coal or oil-fuelled power station might kill dozens. An accident at a nuclear plant might, over time, kill thousands. Hazards associated with nuclear power generation

BUILDINGS AND INFRASTRUCTURE

Fukushima Daiichi was built on a site that stood about 33 feet (10 metres) above sea level. The site was bounded by a 33 foot-high sea wall. The 11 March tsunami overtopped this wall with ease. The inadequacy of Fukushima Daiichi's sea wall created a latent error or resident pathogen. According to the World Nuclear Association (2016): 'The tsunami countermeasures taken when Fukushima Daiichi was designed and sited in the 1960s were considered acceptable in relation to the scientific knowledge then, with low recorded run-up heights for that particular coastline'. In 1896, an earthquake generated a tsunami with a run-up height of 125 feet (38 metres). Given this fact, it is difficult to understand how Fukushima's designers concluded that a 33 foot-high sea wall would offer adequate protection *in perpetuity*.

Japan's coastline has been threatened by tsunamis throughout history. An earthquake in 1993 triggered a tsunami that ranged in height from 33 to 66 feet (10 to 20 metres). The tsunami killed over 200 people. This was the largest tsunami-related death-toll in fifty years (Titov and Synolakis 1997). Other warnings included:

- eight tsunamis in the century prior to Fukushima with maximum amplitudes at origin above 33 feet (10 metres);
- a tsunami in 1983 with a maximum amplitude at origin of 48 feet (14.5 metres);
- International Atomic Energy Authority (IAEA) guidelines that encouraged designers to take account of worst-case tsunami risk (World Nuclear Association 2016).

Warnings and exhortations had little effect on the outlook and behaviour of the key players in Japan's nuclear programme. Neither the TEPCO nor Japan's nuclear regulator, the Nuclear and Industrial Safety Agency (NISA), acted on the IAEA's guidelines: 'Discussion was ongoing, but action minimal' (World Nuclear Association 2016). With reference to Toft's (Toft 1992; Toft and Reynolds 1997) theory of mitigation, learning was passive, not active. Although aware of the risk, the TEPCO and the NISA chose inaction over action. The sea wall could have been made more substantial. It was not. The emergency generators that were vulnerable to inundation could have been moved to higher ground (where the three new generators had been sited). They were not. Switching gear could have been proofed against inundation. It was not. '[The] NISA continued to allow the Fukushima plant to operate without sufficient countermeasures' says the World Nuclear Association (2016). With reference to Reason's (1990, 1997, 2013) theory of system failure, the inaction of the TEPCO and the NISA created latent errors that increased the likelihood of catastrophic failure (because latent errors have the potential to weaken system defences, and because they can be transformed by circumstance into active errors).

PROCESSES AND PROCEDURES

When the Fukushima disaster occurred, Japan's nuclear power industry was regulated by the NISA, a specialist unit within the Ministry of Economy, Trade and Industry (METI). The NISA performed two roles. First, promoting the nuclear industry. Secondly, regulating it. Potentially these roles were in conflict, not least because nuclear power generation was considered a building block of national security.

The NISA was overseen by the Nuclear Safety Commission (NSC) (a safety agency) and the Atomic Energy Commission (AEC) (responsible for strategy). Both the NISA and the NSC were institutionally compromised: The NISA because of its location within the nuclear-industry-promoting Ministry of Economy, Trade and Industry, and the NSC because of its jingoism, technophilia and pro-industry agenda. The Chair of the Nuclear Safety Commission at the time of the disaster alleges that the Commission had 'succumbed to a blind belief in the country's technical prowess and failed to thoroughly assess the risks of building nuclear reactors in an earthquake-prone country' (Madarame cited in Kingston 2012: 2).

Given the Japanese government's linking of nuclear power with national security, it is hardly surprising that a commission embedded in the Cabinet of Japan would seek to de-emphasise operational risk. As to the NSC's organisational culture, priorities and outputs, the Chair of the Commission at the time of the disaster observed: 'We ended up wasting our time looking for excuses that [more stringent standards] are not needed in Japan' (Madarame cited in Kingston 2012: 2).

Of course, any country that links nuclear energy with national security is likely to view it through rose-tinted spectacles. This was certainly the case in Britain, where successive governments framed nuclear power generation as a means of resurrecting the country's war-battered, diminished economy. The Second World War had effectively bankrupted Britain. The country succumbed to 'recurrent financial crises' under the leadership of an 'exhausted' government (Perryman 2006: 9). It was claimed that nuclear power would produce electricity 'too cheap to meter' (a boast originally made in 1954 by US Atomic Energy Commission Chairman Lewis Strauss). Nuclear power generation was presented as a saviour. Risks (when acknowledged) were framed as acceptable. A deferential, technophiliac society tired of shortages and yearning for a better life acquiesced.

With reference to Weick, Sutcliffe and Obstfeld's (1999) theory of decisionmaking, the NSC's decisionmaking was *mindless* in its consideration of risks and hazards. In the field of design, there is a positive relationship between mindlessness and the prevalence of latent errors/resident pathogens. Regarding the operation of complex systems, mindlessness and vulnerability are positively correlated (see Figure 12).

With reference to Rasmussen's (1997, 1999) theory of safety migration, the more the NISA and the NSC obfuscated, the more Japan's nuclear power stations operated at the edge of the safety envelope. Kingston (2016: 1) alleges regulatory capture (see Stigler (1971) for a definition of regulatory capture) of the NISA and the NSC by Japan's utilities and their backers: 'In Japan, nuclear regulators have … long been regulating in the interests of the regulated'. Further, in the matter of the planning and execution of Japan's nuclear power programme, Kingston (2016) alleges groupthink (see Janis (1972) for a definition of groupthink) on the part of politicians, civil servants and industrialists.

In September 2012, the government introduced a new safety regime. In an effort to eliminate – or at least reduce – conflicts of interest, the newly

Systems-thinking in practice

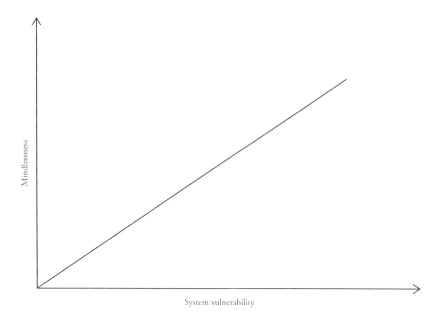

Figure 12. Mindlessness in respect of design and operation increases system vulnerability.

created Nuclear Regulation Authority (NRA) was embedded within the Ministry of the Environment. The Chair of the new agency observed: '[P]ublic trust on nuclear regulation has been completely lost' (Tanaka cited in World Nuclear News 2012).

CULTURE

Culture is '"the way we do things *here*" – the natural and unquestioned mode of viewing the world' (Helmreich and Merritt 2001: 1). According to Kingston (2012), cultural traits helped Japan realise its nuclear ambitions. Helpful traits included:

- murahachibu, which, in the context of Japan's nuclear programme, ensured the social exclusion of those who questioned Japan's nuclear policies. Murahachibu is a culturally articulated groupthink;

- amakudari, which, in the context of Japan's nuclear programme, saw bureaucrats whose decisions supported the nuclear programme rewarded with sinecures;
- corporatism, which, in the context of Japan's nuclear programme, saw parties who stood to gain from the programme (trade unions, construction companies, research laboratories, financiers, investors, academics, etc.) co-operate for the duration.

The National Diet of Japan's (2012) report into the disaster identified another problematic trait – civil servants' defensiveness. '[T]he first duty of any individual bureaucrat is to defend the interests of his organisation' observed The National Diet of Japan (2012: 9). The report claimed that defensiveness influenced civil servants' 'collective mindset' (The National Diet of Japan 2012: 9). Janis (1972) notes how groupthink induces blind loyalty to arguments and causes. When groupthink obtains, arguments and facts that challenge received wisdom are either ignored or dismissed out of hand.

Japan's nuclear programme had been propelled by a powerful actor-network since the 1950s. Viewed through the prism of actor-network theory, it consisted of:

- the utilities;
- political parties;
- trade unions;
- construction companies;
- government agencies;
- research laboratories;
- academia;
- sympathetic media outlets;
- the post-war national reconstruction imperative;
- technological exuberance;
- national cultural traits;
- energy poverty.

The International Atomic Energy Agency (2015: 63) claimed in its report into the disaster that Japan's regulatory agencies, in respect of their governance of the nuclear power industry, had failed to meet required standards: 'The regulations, guidelines and procedures in place at the time of the accident were not fully in line with international practice in some key areas, most notably in relation to periodic safety reviews, re-evaluation of hazards, severe accident management and safety culture'.

The National Diet of Japan (2016: 43), in its report into the disaster, drew attention to the consequences for public safety of collusion between those tasked to deliver Japan's nuclear power programme: '[There was] a cosy relationship between the operators, the regulators and academic scholars that can only be described as totally inappropriate. ... [T]he regulators and the operators prioritised the interests of their organisations over the public's safety, and decided that Japanese nuclear power plant reactor operations "will not be stopped". ... The Commission found that the ... relationship [between Japan's nuclear power industry and those responsible for regulating it] lacked independence and transparency, and was far from being a "safety culture". In fact, it was a typical example of "regulatory capture", in which the oversight of the industry by regulators effectively ceases'. Viewed through the prism of actor-network theory, the body responsible for representing the industry, the Federation of Electric Power Companies (FEPC), in concert with other nuclear industry actants, had translated (won over) the body charged with regulating it, the Nuclear and Industrial Safety Agency (NISA).

PEOPLE

National cultures that value and nurture non-conformity are mindful, creative cultures. National cultures that value and nurture conformity are mindless, uncreative cultures. Pre-War Britain possessed a mindful national culture. Pre-war Germany and Japan possessed mindless cultures. Nazi art was memorable for being unmemorable: Bland, conformist propaganda.

Mindfulness and safety are positively linked (see Figure 13). Regarding the relationship between mindfulness and safety, Mason (2004: 140) observes: '[T]he highly reliable organisation (HRO) model [is] based on the concept of mindfulness. [Highly reliable] organisations are constantly aware of the possibility of failure, appreciate the complexity of the world they face, concentrate on day-to-day operations and the little things, respond quickly to incipient problems and accord deep respect to the expertise of their members. They value knowledge and expertise highly [and] communicate openly and transparently'. Mindlessness – at both the societal and organisational level – poses a threat to operational safety.

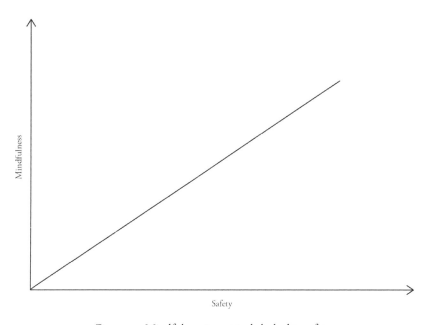

Figure 13. Mindfulness is positively linked to safety.

The evidence suggests that Japanese society is located at the wrong end of the mindfulness-mindlessness spectrum. Helmreich and Merritt (2001) noted differences in outlook and behaviour between pilots socialised in the west, and pilots socialised in the east. Pilots who had been socialised (that is, raised) in the United States were non-conformist. Pilots who had been socialised in Korea and Japan were conformist: 'In line with other cross-cultural communication research … Japanese and Korean pilots were notable for showing greater concern for harmony in the cockpit. … [They were] less willing to disagree openly, less willing to speak up if a problem is detected' (Helmreich and Merritt 2001: 69–70). Yamamori and Mito (1993) claim that to preserve workplace harmony, Japanese workers focus their competitive instincts externally rather than internally.

The final word on the human factor in the Fukushima disaster belongs to Japan's legislature, the National Diet. The Diet suggested that the migration of cultural traits such as conformity and deference into the world of business had created conditions that invited incident and accident: 'What must be admitted – very painfully – is that this was a disaster "Made in Japan". Its fundamental causes are to be found in the ingrained conventions of Japanese culture: our reflexive obedience; our reluctance to question authority; our devotion to "sticking with the program"; our groupism; and our insularity. Had other Japanese been in the shoes of those who bear responsibility for this accident, the result may well have been the same' (The National Diet of Japan 2016: 9).

CONCLUSIONS

Viewed through the lens of actor-network theory, the origins of the 2011 Fukushima nuclear disaster were natural, technical and social. Fukushima was a system accident spawned by a syndrome of weaknesses, dysfunctions and errors-of-judgement within and without Japan's nuclear power industry. The nuclear industry's resident pathogens transformed a natural event – the Tōhoku magnitude 9.0–9.1 (Mw) undersea megathrust earthquake – into a man-made disaster that killed, injured and contaminated (see Figure 14).

Despite the damage suffered by the Fukushima complex, there were no on-site deaths. Unfortunately, some thirty people died in the subsequent evacuation. It is believed that the long-term risk to health from radioactive contamination is minimal. While the images of explosions within the complex left a lasting impression, the real damage was done by the tsunami: 'The tsunami inundated about 560 sq. km [216 sq. miles] and resulted in a human death toll of about 19,000 and much damage to coastal ports and towns, with over a million buildings destroyed or partly collapsed' (World Nuclear Association 2016: 1). The cost of the clean-up has yet to be calculated.

Without doubt, the bravest and most profound analysis of the origins of the Fukushima disaster was produced by The National Diet of Japan. The Diet's capacity for reflexivity and self-effacement produced numerous insights. For example, that Japanese custom and habit helped create the conditions for disaster. As the Diet elegantly put it: '[T]his was a disaster

"Made in Japan"'. The Diet's most devastating claim was that substitution may not have prevented the disaster: 'Had other Japanese been in the shoes of those who bear responsibility for this accident, the result may well have been the same'. Fukushima suggests that some cultures may be more accident-prone than others. Safety correlates with mindfulness. Other things being equal, the more one tests one's assumptions, the lower the likelihood of mishap. Conformity, deference, servility, obsequiousness and brown-nosing make mishap more likely. Other things being equal, societies that encourage introspection and debate are safer than those that inhibit or proscribe it. The rich and powerful are less likely to misbehave if they fear their misdemeanours might be exposed in the debating chamber or in the press. The Fourth Estate is a foundation-stone of public safety. For example, in the 1970s, a campaign by *The Sunday Times* helped move the Thalidomide scandal up the political agenda. China's poor mines safety record has been attributed to political corruption, lack of trade union organising and the absence of a free press. Because it suppresses freedom of thought and action, totalitarianism foments mishap.

Figure 14. International Atomic Energy Agency experts walk the Fukushima Daiichi site post-disaster (Wikimedia Commons 2018g).

Offshore oil and gas production accident investigation

Winning oil and gas from the earth is not without risk. This is especially the case in the icy and turbulent waters of the North Sea. The rewards, however, are great. The Piper Alpha oil-and-gas production platform, operated by Los Angeles-based Occidental Petroleum, was once Britain's biggest platform, 'bringing more than 300,000 barrels of crude a day – 10% of the country's total from below the seabed 125 miles north-east of Aberdeen' (*The Guardian* 2013). Oil was discovered under the Piper field in 1973. By 1976 it was being pumped. The boom was short-lived. A falling oil price hit the industry and its workers hard: 'Along with other oil companies, Occidental had massively scaled back spending as the price of oil had plunged from more than $30 per barrel to $8 in the 1980s compared to today's level of more than $100' (*The Guardian* 2013).

On the night of 6 July 1988, a series of explosions destroyed the Piper Alpha oil-and-gas production platform, killing 167 workers. Lord William Cullen's public inquiry into the Piper Alpha disaster was informed by systems-thinking. While no criminal charges were brought against Occidental, each of the inquiry's 106 recommendations was accepted by the industry. Some of the Cullen inquiry's recommendations touched on the way the industry was managed by the State.

The disaster is estimated to have cost £2.4 billion ($3.4 billion). The human cost was very great. Men were drowned or burned alive. Some survivors and rescuers were traumatised. Television coverage magnified the horror. One survivor recalled: 'There were twenty lifejackets where I was for forty men. We gave the older men preference. The heat was so intense we had to get into the water. Me and five other men clung onto a ladder but four of them had to give up in the end. They let go and just drifted away. Me and another bloke managed to hang on for more than half an hour, when a rescue boat turned up but couldn't get us on board because of the danger of falling debris. We kicked our boots off and managed to swim to it. There were bodies all around us in the water badly burned, with strips of skin coming off them' (Niven cited in Hall 1995: 51).

A rescuer recalled: 'We went right in under the rig [in our inflatable boat] and picked up four crewmen. We went back in again to pick up another two, but then there was this colossal explosion and the boat was engulfed in flames. My hard hat and jacket just melted in the heat. I never saw the two we picked up again – they were gone' (Letham cited in Hall 1995: 49). A helicopter pilot was moved to praise rescuers' bravery: 'Time and time again they went back into that heat. We could feel the heat from a mile away as we flew in' (Porter cited in Hall 1995: 49).

Piper Alpha

The North Sea is unforgiving of any weakness, slip or lapse (see Figure 15). In 1980, the Alexander Kielland, a four-year-old, 10,000 ton accommodation platform, overturned in heavy seas (whipped up by 130 mph winds), sending 123 of the 213 workers on board to the bottom. Immediate causes included a structural failure caused by metal-fatigue, progressive failure of proximate components and inadequate contingency planning: 'The 14 minutes between initial failure of the leg and the rig's eventual capsize left a window in which most of the personnel on board could have escaped, given a more effective command structure. But it would seem that no one took charge on the night' (Officer of the Watch 2013).

Proximate causes included deficiencies in rules and codes pertaining to design and manufacture, errors made during manufacture and inadequate provision of life boats and survival equipment. Similar installations were modified post-disaster. The Kielland capsize demonstrated the vulnerability of even the most substantial structures. In hindsight, Kielland was a warning. Lagadec (1982: 495) observes: 'The disaster must not be seen like a meteorite that falls out of the sky on an innocent world; the disaster, most often, is anticipated, and on multiple occasions'.

In 1912 the RMS *Titanic* hit an iceberg and sank. *Titanic*'s twenty lifeboats could accommodate just under 1,200. There were over 2,200 on board when the vessel was fatally holed below the waterline. It appears that those who designed the Alexander Kielland had forgotten the lessons of the RMS *Titanic*.

There were no further large-scale disasters in the North Sea until the Piper Alpha disaster of 1988, the origins of which are described here with reference to Challenger, Clegg and Robinson's (2009) simple but illuminating topographical model of system behaviour:

Systems-thinking in practice 61

Figure 15. Rigs in the Cromarty gas field photographed in untypical weather. Turbulence in the Middle East in the 1960s and 1970s incentivised the rapid exploitation of local oil and gas reserves, even in the unforgiving North Sea (Wikimedia Commons 2018h).

GOALS

The goals of those involved in the extraction of oil and gas from the North Sea in the 1970s and 1980s were many and varied. While there was concern that safety margins were too thin (Whyte 2006), it must be said that the workforce was, to a degree, complicit in corner-cutting. Safety was far from the *sole* concern of workers. Many saw offshore work as an opportunity to make a killing – a means of setting themselves and their loved ones up for life. Hall (1995: 47) observes: 'There was big money to be made working the oil rigs. ... A new breed of frontiersman, earning upwards of £1,000 a week, was lured to the steel outposts. ... The work was hard and dirty, completed in tours of duty. Like soldiers, the men who worked the rigs put up with long spells away from home ... but when they went ashore for leave it was with their pockets bulging with cash and the knowledge that they had between a fortnight and a month before going back'. Understandably, the trade union movement claimed that offshore workers had a robust and abiding interest in safety, in terms and conditions and in the long-term prospects of the industry. The reality was somewhat different. Yes, workers were concerned about safety. But they were also concerned about exploiting opportunities to improve their economic standing and life-chances. Some offshore workers were died-in-the-wool 'union men'. Some were 'company men'. Some had divided loyalties. Things are never as black-and-white as those with an agenda would have us believe (trade union leaders, for example).

Britain was a troubled land in the 1970s. Hit by oil crises, industrial unrest, under-investment, industrial decline, slow growth, bad choices (Concorde, an expensive white elephant, was funded, while the promising British Aircraft Corporation TSR2 strike aircraft project was terminated), racial tension, political polarisation, terrorism and a loss of confidence, the country had earned the soubriquet 'the sick man of Europe': 'A poor growth record since the second world war combined with terrible industrial relations (29m days lost to strikes in 1979) to make many ask the question "Is Britain governable?"' (*The Economist* 2017). So desperate did Britain's political class become that it begged the European Economic Community to let it join. Many saw membership as a means of introducing change to a beleaguered economy (*The Economist* 2017). If you were lucky enough to have a well-paid job in the 1970s, you would probably have done everything in your power to hold on to it.

Men laboured long and hard in an industry shaped by corporate greed and by governments' prioritisation of the national interest over other concerns. The *primary* responsibility of any government is to secure its citizens. Securing the population trumps every other concern, including reducing the national debt, subsidising health care, paying for schools and protecting the environment. Rocked by adverse events at home (for example, the miners strikes of 1972 and 1974 that saw rolling blackouts and a three-day working week) and abroad (for example, the 1970s oil-shocks), governments of every persuasion saw the revenues from North Sea oil as a means of rebuilding the British economy.

Systems-thinking in practice

The Thatcher government saw the revenues from North Sea oil as a means of underwriting a fundamental restructuring of the British economy. Oil revenues helped pay the social security benefits of those who lost their jobs to de-industrialisation. As Whyte (2006: 185) explains: 'There is reason to believe that the U.K. government's pioneering strategy of neo-liberal economic reform could not have proceeded without oil. Alan Clarke, a veteran member of several Thatcher governments, has since argued that, "without the revenue from oil, there could have been no Thatcherism"'.

Given its pivotal role in saving the Sick Man of Europe from economic collapse, it is unsurprising that governments did everything in their power to encourage the oil giants to extract as much oil and gas from the North Sea as quickly and as cheaply as possible. The regulatory environment created by government was conducive to risk-taking and profit-making and hostile to trade-union-inspired restrictive practices. The mores and practices of Britain's North Sea oil and gas industry reflected those of neo-liberal market economics ... of Reaganomics and Thatcherism.

Reaganomics and Thatcherism influenced social relations. The North Sea oil worker was a factor of production, a commodity to be hired and fired as required. Short and very short-term contracts became the norm:

> [W]orkers were usually employed on short-term contracts that often lasted no longer than a few weeks or months, and contained few contractual rights. Trade union organisation was virtually unheard of. Authoritarian, often thuggish, management styles ensured that those who were found to have trade union sympathies, who complained too much about working conditions, or who vocally expressed concerns about safety, were run off the platform ... Eighty-three percent of workers on board Piper Alpha on the night of the disaster were sub-contracted ... They therefore hardly enjoyed a stable vantage point from which to present their concerns to management on questions of safety. The marginalisation of workers' expertise and knowledge of safety ... was to have catastrophic consequences. (Whyte 2006: 184)

There is a negative correlation between staff churn and safety because it diminishes organisational memory: 'A key pre-requisite for experience-based learning is the extent to which institutional memory is cultivated and accessible to participating actors. ... Unfortunately ... many ... kinds of organisations tend to be weak in this area. ... Valuable competence and stores of experience are routinely lost through staff attrition. As a result, organisations forget' (Stern 2008: 288). The politically sanctioned casualisation of the North Sea oil and gas industry's labour force increased the likelihood of incident, accident and near-miss.

TECHNOLOGY

Machines designed for use on dry land proved unreliable. Unable to withstand the adverse weather conditions, they broke down. Twenty-four-hour working (supported by a two-shift system) and infrequent maintenance increased the risk of breakdown. According to Whyte (2006), the 1985 collapse of the OPEC quota system and determination of those involved in the North Sea to defend profits saw budgets cut by up to 40 per cent. Offshore workers were laid off. Wages were cut. Fear of redundancy stalked the rigs. Maintenance budgets and schedules were pared back. Offshore facilities were operated at the edge of the safety envelope. Rasmussen (1997) notes how systems subject to economic pressure may drift into danger. He describes as 'natural' under-pressure systems' 'migration toward the boundary of functionally-acceptable performance' (Rasmussen 1997: 189).

Piper Alpha's workforce was under pressure from platform operator Occidental Petroleum which, in turn, was under pressure from a falling oil price, a get-rich-quick industry culture, impatient shareholders and a British government determined to maximise tax revenues. The social, economic and political climate induced safety migration.

BUILDINGS AND INFRASTRUCTURE

Evolving operational requirements required that the North Sea oil and gas industry's infrastructure be adapted and modified. The accommodation platform Alexander Kielland started life as a drilling rig. Following the UK government's 1978 adoption of a gas-conservation policy, Piper Alpha, designed to process crude oil, was converted to process both oil and gas. According to the National Aeronautics and Space Administration (2013: 1–2), Occidental's reactive patching of Piper Alpha (see Weir (1996) for a definition of reactive patching), that saw a gas-compression module (GCM) located adjacent to the platform's control room, violated the safe design concept: 'After modification, Piper Alpha processed gas. ... Piper Alpha additionally served as a hub, connecting the gas lines of two other Piper field platforms. ... [T]he modifications to Piper Alpha broke from the safe design concept that separated hazardous and sensitive areas of the platforms'.

Occidental's upgrade and operating practices created risk: 'The original modules on the structure were carefully located, with the staff quarters kept well away from the most dangerous production parts of the platform. But this safety feature was diluted when the gas compression units were installed next to the central control room. Further dangers arose when Occidental decided to keep the platform producing oil and gas as it set about a series of construction, maintenance and upgrade works' (Macalister 2013). Viewed through the prism of Perrow's (1984) normal accident theory (NAT), the installation of the GCM increased the platform's interactive complexity. It made failure more likely.

The UK government's gas-conservation policy required a response. Occidental could have adapted Piper Alpha in such a way that safety margins were maintained (or improved). Instead, Occidental chose the line of least resistance. It located the new, mandated GCM next to Piper's control room. In doing so, the company compromised the safety of the Piper Alpha platform and its workers. Occidental's failure to upgrade the platform's fire walls introduced another latent error or resident pathogen into the system. When the GCM exploded on the night of 6 July 1988, the walls buckled: 'Firewalls designed to withstand burning oil, crumbled under the overpressure from the detonating gas' (National Aeronautics and Space Administration 2013: 2).

Losing the control room to the explosion and fire sealed the platform's fate:

> As Piper Alpha was never designed to pipe gas as well as oil, the new gas pipes were installed wherever there was room. Unfortunately, the lines affected by the explosion

were located next to the control room and the resultant explosion rendered essential disaster management impossible. Although the emergency stop button was pressed, no alarm was sounded throughout the rig and the fire suppression system was never activated. The deluge system that pumped water from the sea into sprinklers located throughout the station could only be activated from the control room, as it was switched off to allow divers to work below the rig. The fire spread unchallenged and eventually ruptured a gas riser causing a massive explosion that engulfed the whole platform. (Bell 2006: 41)

PROCESSES AND PROCEDURES

Under the circumstances of 6 July 1988, some safety procedures served to increase rather than reduce the risks to Piper Alpha's workforce. For example, it was standard practice for the platform's deluge system to be switched to manual operation when divers were working close to the system's powerful intakes. Loss of the control room on the night of 6 July meant the system could not be activated. Prior to the disaster, it had been suggested that the deluge system should be deactivated only if Piper's divers were working close to the intakes. Management ignored the suggestion. The switching policy stood: 'Earlier audit recommendations suggested that pumps remain in automatic mode if divers were not working in the vicinity of the intakes, but this recommendation was never implemented' (National Aeronautics and Space Administration 2013: 3).

According to Merton (1936), purposive social action ('action which involves motives') can have intended (expected), and unintended (unexpected) consequences (which he terms 'functions'). Manifest functions are the consequences we expect. Latent functions are those we do not. There are two types of latent function: those that support the original intent, and those that work against it ('latent dysfunctions'). Occidental's deluge system policy (that required the switching of the system to manual operation when divers were in the water), although well intentioned, produced a latent dysfunction – specifically the need to manually activate the system in an emergency. Little or no thought seems to have been given to the problem of activating the deluge system when Piper Alpha's control room was operating in degraded mode or was out of action. Occidental's deluge system policy degraded system resilience. It created a latent error or resident pathogen in platform operation.

Under the circumstances of 6 July 1988, another of Occidental's policies – that during an emergency, workers should assemble in the accommodation module and await instruction – produced a latent dysfunction. Given that Piper's fire walls were not designed to withstand blast-induced overpressures, and that the control room had been destroyed by explosions and a furious conflagration (fuelled by oil and gas), the dictum that workers should assemble in the accommodation module was, effectively, a death sentence: '[T]he accommodation area was in the direct path of the ... fireball, and many staff members, who followed the emergency procedures, died' (Bell 2006: 42).

According to Reason (2013: 71), those who ignored Occidental's 'stay put' policy stood the best chance of getting off the platform: 'Those – the majority – who died, followed the procedures; those – the few – who lived, did not. ... The safety regulations required all personnel to assemble by the accommodation in the event of an emergency. Unhappily, this position put them in the line of a plume of smoke and flames'.

Under the circumstances that obtained on the night of 6 July, Occidental's accommodation module assembly policy, while well-intentioned, reduced workers' chances of survival. The closing-down of escape options by the fire made the situation worse: 'The scale of the fire and smoke made the planned evacuation options of helicopter and lifeboat impractical and the survivors escaped to the sea by jumping or using ropes or hoses' (Royal Academy of Engineering 2005: 68). Had managers introduced greater flexibility into Piper Alpha's evacuation drills, more workers might have survived. Emergency procedures that allow for on-the-spot initiative may produce better results than those that do not, especially when events unfold in unpredictable ways. Mindfulness (Weick, Sutcliffe and Obstfeld 1999) can save lives.

Workers can be an asset. In Hollnagel, Wears and Braithwaite's (2015: 27) Safety II portrayal of the worker, 'Humans are seen as a resource necessary for system flexibility and resilience [providing] flexible solutions to many potential problems'. Hollnagel, Wears and Braithwaite (2015: 17) promote the human-factor: 'People who contribute ... intelligent adjustments are ... an asset'. Evidence for humans' positive contribution to the management of complex systems includes Soviet officer Stanislav Petrov's actions (actually, considered inactions) that helped prevent a nuclear exchange during NATO's 1983 Able Archer exercise.

Hollnagel, Wears and Braithwaite's (2015) Safety II conception of the worker as an asset resonates with high reliability organisation (HRO) (Roberts 1990; LaPorte and Consolini 1991; Christianson, Sutcliff, Miller and Iwashyna 2011) theory's conception of the worker as, potentially, a source of problem-solving ideas and skills.

The immediate cause of the disaster was the failure of Piper Alpha's Permission-To-Work system. *Absentia* the outgoing shift briefing the incoming shift, the new shift got up to speed by reviewing manual records. On the night of 6 July, this system failed:

> During the day shift, staff had removed a pressure safety valve from [a] pump. The valve had been replaced with a blank flange. The blank fit was not tight enough, and when unaware night-shift staff then attempted to restart the pump, condensate (light oil) leaked through, causing [an] explosion. ... The flange placed over the valve was a temporary measure and was inadequate to deal with operational gas pressures. Staff in the control room were unaware of any work that had been done in this area, as the Permission-To-Work ... certificate had been lost. They subsequently activated the valve and caused a massive system-pressure overload. Condensate escaped from the piping and ignited causing a large explosion. (Bell 2006: 41)

Had Occidental required the outgoing shift to brief the incoming shift face-to-face, the disaster might have been avoided. The absence of such a requirement suggests a lax safety culture.

CULTURE

The Piper Alpha disaster originated in the social, economic and political landscape of the 1960s and 1970s. When oil was cheap and the world in step with Western values, there was little incentive to improve fuel efficiency, develop alternative sources of energy (such as solar, wind, wave or hydro-electric power) or look for new, more secure reserves. Cheap oil induced lazy thinking and complacency and subsidised costly diversions such as the space race, supersonic flight and the Vietnam war (see Figure 16). The 1973 oil embargo focused Western minds. The quadrupling of the price of crude (from $3 a barrel in 1973 to $12 a barrel in 1974) impacted the UK socially, economically and politically.

Figure 16. In the same way that gas-guzzlers were a product of the era of cheap oil, the Piper Alpha disaster was in part a product of the oil-shocks of the 1970s, Reaganomics and Thatcherism (Wikimedia Commons 2018i).

The oil-shocks of the 1970s magnified the UK's social, economic and political travails. Wracked by inflation (which stood at almost 25 per cent), class antagonism, racial tension, industrial strife, falling profitability, collapsing share prices, low productivity, chronic under-investment, poor strategic choices (the TSR2 project had been abandoned and Concorde was a costly flop), Irish Republican terrorism and, for a while, a three-day working week accompanied by rolling black-outs, the British economy faced a bleak future (Macalister 2011). According to Hall and Jacques (1989: 29), during the 1970s, Britain experienced 'a tumultuous economic, social and political crisis … created by long-run weaknesses'. The country needed a miracle. That miracle was North Sea oil.

Given the importance of North Sea oil to Britain's prospects, it is hardly surprising that governments did everything in their power to support the industry. Whyte (2006: 184–185) notes: 'This was a period in which productive oil capital enjoyed a structural advantage in relation to policy development. ... It is in the consolidation of oil capital's hegemonic position that the genesis of the regulatory failure in the pre-Piper Alpha period is to be found'.

Desperate governments make bad choices. Regarding North Sea safety, governments' bad choices included deference to oil companies' *modus operandi* (aloof, top-down management, aggressive cost-cutting and antipathy toward trade union organising) and a light-touch system of regulation that saw responsibility for offshore safety vested in the department of state responsible for nurturing the industry – the Department of Energy (DoE). The Department of Energy's institutional focus on nurturing North Sea oil helped create the conditions for the Piper Alpha disaster.

To a degree, the 6 July 1988 Piper Alpha disaster originated in an institutional conflict of interest borne of a *conscious political decision* to prioritise output (Whyte 2006; National Aeronautics and Space Administration 2013). In July 1988, the Department of Energy had at its disposal seven inspectors to audit every installation in the North Sea (Whyte 2006). Hollnagel's (2004, 2013) theory of an Efficiency-Thoroughness Trade Off (ETTO) suggests a negative relationship between economic performance and safety performance. That is, the more an industry chases profit, the closer to the outer edges of the safety envelope it operates.

Economically, governments' oil-first policy was a success. By the end of the 1970s, the British economy was recovering. The Sick Man of Europe was still sick, but he wasn't going to die: '[By 1978] the balance of payments was ... back in surplus by £1,000m; North Sea oil revenue was flowing in; inflation was falling fast; the index of industrial production rose steadily; even unemployment showed a downward turn' (Morgan 1987: 321). By the end of the 1980s, North Sea oil revenue had supported a decisive and irreversible restructuring of Britain's economy. Britain had spawned a late industrial, get-rich-quick economy supported largely by its service sector. Morally – with specific reference to Piper Alpha – governments' oil-first policy was a failure. It killed and maimed. It left families without breadwinners.

As mentioned, the North Sea oil industry spawned a brutish culture. Installation managers, intolerant of dissent (Whyte 2006; Beck and Drennan 2000), were reluctant to listen to workers. Beck and Drennan (2000: 1) allege an 'authoritarian labour relations regime, in which workers'

Systems-thinking in practice 71

voices were ignored'. The energy companies were far from being high reliability organisations.

On Piper Alpha, corporate intolerance and selective deafness saw managers dismiss the divers' argument that the policy of switching the deluge system to manual operation whenever a diver was in the water was too coarse. The divers argued that the policy jeopardised the platform's safety. According to the National Aeronautics and Space Administration (2013: 3) 'divers did not see significant risk unless they were working closer than 10 to 15 feet from any of the intakes'.

Divers' opinions were ignored. The policy of switching the deluge system to manual operation remained in place. The policy contributed to the 6 July 1988 disaster.

This episode suggests that Occidental's organisational culture was not a learning culture. In a learning culture, an organisation 'is committed to learn safety lessons; communicates them to colleagues; remembers them over time' (Carthey and Clarke 2010: 10). The episode suggests also that Occidental's organisational culture was mindless, that is, unreflexive (see Weick, Sutcliffe and Obstfeld (1999) for a definition of mindlessness).

PEOPLE

According to Carthey and Clarke (2010), a positive safety culture has five elements. These are described in Table 1.

Table 1. Elements of a positive safety culture (Carthey and Clarke 2010: 10; apart from the removal of references to healthcare, the table is reproduced exactly as drafted by the authors).

Open culture	Staff feel comfortable discussing ... safety incidents and raising safety issues with both colleagues and senior managers
Just culture	Staff ... are treated fairly, with empathy and consideration when they have been involved in a ... safety incident or have raised a safety issue

(*Continued*)

Table 1. (*Continued*)

Reporting culture	Staff have confidence in the local incident reporting system and use it to notify ... managers of incidents that are occurring, including near-misses Barriers to incident reporting have been identified and removed: • staff are not blamed and punished when they report incidents • they receive constructive feedback after submitting an incident report • the reporting process itself is easy
Learning culture	The organisation: • is committed to learn safety lessons • communicates them to colleagues • remembers them over time
Informed culture	The organisation has learnt from past experience and has the ability to identify and mitigate future incidents because it: • learns from events that have already happened (for example, incident reports and investigations)

The Piper Alpha disaster occurred in the context of an organisational culture that was neither open, nor just, nor informed. It occurred in the context of a culture that was neither a reporting culture nor a learning culture. There was little incentive for workers, who were generally treated with disdain, to vocalise safety concerns. Indeed, according to Whyte (2006: 184), those who did risked being black-listed: '[Workers] who complained too much about working conditions, or who vocally expressed concerns about safety, were run off the platform (in industry parlance "NRB'd" [not required back]) as a matter of routine'.

Piper's dysfunctional culture meant that learning was sometimes passive. Nine months before the platform blew up a worker had been killed. Contributory factors included a confused shift-handover and deficiencies in the Permission-To-Work system. According to Cullen (cited in *Finding Petroleum* 2013): 'After that incident, management took some steps, but they were not followed through'. Occidental's failure to rectify systemic weaknesses (for example, in respect of the Permission-To-Work system and handovers) was typical of the industry as a whole, which tended to focus on slips, trips and falls rather than on ensuring the integrative, holistic integrity of complex, risk-laden socio-technical systems and effectiveness of safety policy and procedure. There was little joined-up thinking: 'A theme of accidents over the past decade has been too much emphasis on personal safety (hard hats) and not enough on process safety (the accidents which cause the big disasters)' (*Finding Petroleum* 2013). The *sine qua non* of safety assurance is systems-thinking-informed analysis. The *sine qua non* of systems-thinking-informed analysis is resourcing.

For the North Sea roustabout (platform worker), safety was but one concern amongst many. Other concerns included wage levels, unsocial hours, fitting in, securing follow-on employment, the quality of the food and entertainment, the quality of the accommodation, maintaining relationships with partners and offspring and holding on to one's wage packet. As one worker explained: 'Gambling was onshore and offshore. There were a lot of big card schools with a lot of big money changing hands. I've seen guys going off the platform with thousands of pounds in their pocket – two, three, four, five thousand pounds. It wasn't unknown for a guy who had lost money to be fog-bound offshore for two weeks. In other words, he didn't go home to his wife' (Molloy cited in Gall 2010).

Other interests included alcohol and sex: 'To put it crudely, when we first came to Aberdeen we drank whiskey and shagged [had sex with] pretty girls' (Doubenmier cited in Gall 2010). There were concerns about maintaining relationships with partners and offspring: 'You got one [telephone] call a week, and you had to make the best of that call. At times it could be very frustrating, because you couldn't cover everything, especially if you had young kids. You had to queue to 'phone home, were timed, and had a maximum of six minutes, then [you were] cut off' (Molloy cited in Gall 2010).

The cursory survival training drew little comment: '[T]here wasn't much [training] back then [in the early 1970s]. ... We were thrown in at the deep end and just got on with it. Later on in the 70s, the first offshore survival training started. It consisted of showing us some slides of the work sites and environment offshore, then sitting on an upturned rowing boat in Aberdeen harbour to learn about survival at sea' (Adams cited in Gall 2010).

Frequently, the North Sea roustabout was portrayed as a hapless victim of greedy, mendacious (deceitful) capitalists. Reportage could be hyperbolic. For example, in his 29 July 2016 report on a strike by offshore maintenance workers employed by Wood Group, *Socialist Worker* journalist Raymie Kiernan (2016) referred to workers 'being packed into a chopper like sardines'. Kiernan went on: 'Everything points to disaster, but this is how they get to work'. The journalist omitted to mention that all commercial air operations – including North Sea helicopter operations – are required to meet European Aviation Safety Agency standards. Kiernan claimed that the North Sea oil industry's infrastructure 'is largely past its use-by date'. The journalist omitted to mention that most infrastructure, if properly maintained, can be safely operated beyond its design-life. (Design-life calculations are often based on conservative estimates of durability).

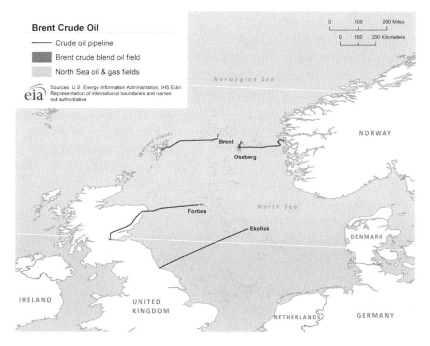

Figure 17. Statistics pertaining to death and injury should be considered against a range of factors, including the scale, complexity and adverse operating conditions of the North Sea oil and gas industry. Rigs and pipelines require constant maintenance (Wikimedia Commons 2018j).

Kiernan claimed that 'Oil and gas barons are prepared to sacrifice lives to line their pockets'. According to the Health and Safety Executive, in 2016, the year in which Kiernan penned his article, 'There was one fatal injury [in the North Sea]. ... There were 20 specified injuries, with a rate of 66 per 100,000 full-time equivalent (FTE) workers' (Health and Safety Executive 2017: 3). There were six fatal injuries between 2006 and 2016. While every death and injury is a stain on the industry, these figures should be considered against the size, complexity and hostile operating environment of the North Sea oil and gas industry (see Figure 17): 'In 2016 ... there were 302 installations in the UKCS [United Kingdom Continental Shelf], of which 261 were operational and 143 were manned. In addition, there is a supporting infrastructure of 14,000 km [8,700 miles] of pipelines connecting installations to beach terminals' (Health and Safety Executive 2017: 3). It is advisable to consider wider situational and systemic factors – that is, the back-story – before putting pen to paper. Every profession, including journalism, can benefit from systems-thinking.

CONCLUSIONS

Viewed through the lens of actor-network theory, the origins of the Piper Alpha disaster were social, economic, political, cultural and systemic. Governments' desire to exploit Britain's last great hope for economic renewal created the trajectory and momentum for the disaster. Politicians of every persuasion were complicit in the disaster. Risk-taking was tolerated at all levels of government, from Downing Street to the Department of Energy's conflicted and under-resourced offshore inspectorate, and in every corner of the industry. Actants implicated in the disaster included:

- Britain's post-war decline. Instigated by the crippling cost of fighting the Second World War, America's abrupt ending of lend-lease and the retreat from Empire, Britain's decline found expression in the Callaghan Government's 1976 negotiation of a £2.3bn loan from the International Monetary Fund;
- instability in the Middle East;
- OPEC;

- falling oil prices (that squeezed profits and investment);
- the greed of national governments;
- the Department of Energy. In 1987, the DoE advised against mandating that North Sea installations produce safety cases. According to Cullen (cited in *Finding Petroleum* 2013), this constituted 'a serious setback for development of the offshore [safety] regime';
- a State-sponsored double standard in safety management. Following the Flixborough chemical plant disaster of 1974 that killed 28 and injured 36, installations were required to produce safety cases. This requirement was never applied to offshore installations. Renn (2008: 52) notes: '[T]he institutional structure of managing and controlling risks is prone to organisational failures and deficits that may increase the actual risk';
- the aloof, top-down management style of corporations such as Occidental Petroleum;
- Piper Alpha's unreflexive, mindless management regime;
- Piper Alpha's weak contingency planning;
- a local Permission-To-Work system peppered with latent errors;
- a shift-change system that tolerated blind handovers (that is, handovers with no face-time between the incoming shift and outgoing shift);
- maintenance workers' complicity in blind handovers;
- Occidental's violation of safe design standards to meet a new government-mandated environmental standard (proscription of flaring and conservation of natural gas);
- Occidental's tolerance of safety migration in the architecture of the Piper Alpha platform (for example, the decision not to install sub-sea isolation valves for Piper's gas risers and the decision not to install blast-proof walls to mitigate the damage caused by overpressures).

On the twenty-fifth anniversary of the Piper Alpha disaster, Lord Cullen, in a keynote address to the Oil and Gas UK Annual Conference, observed:

> Management shortcomings [on Piper] emerged in a variety of forms. For example, there was no clear procedure for shift handover. *The permit to work system* was inadequate. But so far as it went, it *had been habitually departed from*. Training, monitoring and auditing had been poor, the lessons from a previous relevant accident had not been followed-through. Evacuation procedures had not been practised adequately.

Systems-thinking in practice

> There had not been an adequate assessment of the major hazards and methods for controlling them. ... I was conscious that no amount of regulations can make up for deficiencies in the quality of management of safety. That quality depends critically on effective safe leadership at all levels and the *commitment of the whole workforce to give priority to safety*. I saw those factors as intertwined with each other, and together making a positive learning culture and all that entails in the way of values and practises. It is essential to create a corporate atmosphere or culture where safety is understood to be, and accepted as the number one priority [my emphasis]. (Cullen cited in *Finding Petroleum* 2013)

The 1988 Piper Alpha disaster originated in a soup or miasma of factors, some tangible (for example, flawed design), some intangible (for example, pursuit of the national interest), but all contributing in some degree.

Piper Alpha was a societal accident – partly the product of a desire to rectify structural weaknesses in the economy before they damaged it irreversibly. Because of North Sea oil, the country is richer today than it was in the 1960s. As *Before the Oil Ran Out* author Ian Jack (2013) notes: '[M]ore people [own] more things – tumble dryers and deep freezers – than ever before'. The economy has been restructured around the service sector. High-value-adding companies have prospered. Britain's North Sea oil industry has produced corporate tigers. Before the oil boom, Wood Group of Aberdeen ran a fishing fleet. The company capitalised on the North Sea oil-boom. Today, Wood Group's 43,000 employees can be found working with oil and gas exploration companies all over the world (Day 2013). Perhaps the relatives and friends of those who died on Piper can take comfort from the fact that their late husbands, brothers and friends helped transform Britain's fortunes through their endeavours. A victory – of sorts.

Macondo

The multinational British Petroleum (BP) operates across the globe, including in the Gulf of Mexico. On 20 April 2010, BP's Deepwater Horizon drilling rig, operated in association with Transocean and Halliburton, exploded. Eleven workers died. After burning for thirty-six hours, the rig sank (see Figure 18) (British Petroleum 2010). With BP unable to

Figure 18. The rig sank thirty-six hours after the initial explosion (Wikimedia Commons 2018k).

stem the flow of oil from the ruptured wellhead, there were fears of an environmental disaster.

BP had pushed the envelope with the Macondo well: 'The rig was drilling in about 5,000ft (1,525m) of water, pushing the boundaries of deep-water drilling technology' (British Broadcasting Corporation 2010). On 10 May 2010, BP estimated that the disaster had cost the company £233m ($350m). On 14 May, researchers estimated that oil was leaking from the well at a rate of 70,000 barrels a day (with a margin of error of plus or minus 20 per cent).

At a series of Congressional hearings held in mid-May, each of the three companies involved in the Deepwater Horizon drilling operation, BP, Transocean and Halliburton, blamed the other two for the disaster. President Obama referred to the exchanges between BP, Transocean and Halliburton as a 'ridiculous spectacle'. In his 15 June address to the nation, the US President, who by this time had made several visits to the Gulf of Mexico, stated: 'We will make BP pay for the damage their company has caused' (Obama cited in British Broadcasting Corporation 2010). A few days later, the chairman of the House Committee on Energy and Commerce accused BP of complacency prior to the disaster. By early July, the disaster had cost BP £2bn ($3.12bn). At one point, there were 400 oil-skimming ships working to prevent the oil from reaching the Gulf coast. Tropical storm Bonnie compounded BP's problems when it forced the rig that was drilling a relief well to withdraw. Criticised for his handling of the crisis, BP's chief executive was replaced. At the end of July, BP announced it had set aside £20.8 billion ($32.2bn) to cover costs linked to the spill. The well was capped eighty-seven days after the initial explosion. BP admitted to 'causing a spill of national significance' (British Petroleum 2010: 9).

On 23 April 2010, BP launched a fault-tree-analysis-informed internal investigation into the disaster. Terms of reference included:

- '[analysis of] the sequence of pertinent events';
- '[identification of] critical factors and their underlying causes' (British Petroleum 2010: 13).

Running to nearly 200 pages, the report made no mention of organisational culture or safety culture, although it did include the suggestion: 'It may ... be appropriate for BP to consider further work to examine potential systemic issues beyond the immediate cause and system cause scope of this investigation. ... [G]iven the complex and interlinked nature of this accident, it may be appropriate to further consider its broader industry implications' (British Petroleum 2010: 6). The report made no mention of prior significant events, such as the 2005 BP Texas City disaster that cost the lives of fifteen workers (Mogford 2005). In 2017, despite the financial trauma of the Deepwater Horizon disaster, BP's assets were valued at $276.5 billion. In that same year the company employed 74,500 persons.

As a counterpoint to BP's one-dimensional analysis of the origins of the Deepwater Horizon disaster, the following analysis, informed by Challenger, Clegg and Robinson's (2009) model of system behaviour, enlarges the problem space:

GOALS

Hollnagel (2016) argues that the tension between economic performance and safety performance leads to trade-offs: 'In their daily activities, at work or at leisure, people (and organisations) routinely make a choice between being effective and being thorough, since it rarely is possible to be both at the same time. If demands to productivity or performance are high, thoroughness is reduced until the productivity goals are met. If demands to safety are high, efficiency is reduced until the safety goals are met'.

According to the International Civil Aviation Organisation (2002: 1/1), airline operations are characterised by the trading of effectiveness against thoroughness: 'It is inherent to traditional approaches to safety to consider that, in aviation, safety comes first. In line with this, decision-making in aviation operations is considered to be 100 per cent safety-oriented. While highly desirable, this is hardly realistic. *Human decision-making in operational contexts is a compromise between production and safety goals* [my emphasis]'.

Given that the first priority of any capitalistic enterprise is to make money for investors and shareholders (the first priority is not – and never can be – safety), it follows that in capitalistic enterprises the tension between economic performance and safety performance is *invariably* resolved in favour of the former. While most enterprises spend large amounts of money to protect their customers, workforce and plant, the philosophical datum of every investor-led corporation is securing the maximum return on capital. To this end, where possible, costs are externalised: 'Corporations, as institutions that exist to maximize shareholder value, have a built-in compulsion to externalize their costs, i.e., have others pay for them. ... [w]hether through dumping products into the environment, or through increasing safety risks' (Campbell 2013b: 16).

Sloganeering around safety is commonplace. As Cullen (cited in *Finding Petroleum* 2013) observed on the twenty-fifth anniversary of the Piper Alpha disaster: 'Many companies have safety slogans such as absolute safety and zero accidents'. He went on to observe that disasters such as Piper Alpha, Texas City, Buncefield (where, at a fuel depot on the outskirts of London, a lax safety and reporting culture contributed to an explosion and fire that caused serious injury to two persons and injury to a further forty-three), Montara (where a blowout led to the evacuation of a platform and created a significant environmental hazard in the Timor Sea) and Macondo suggest that the reality of a corporation's safety and reporting culture may be out of kilter with its rhetoric about safety. Regarding Buncefield, Cullen (cited in *Finding Petroleum* 2013) observed:

'[V]arious pressures had created a culture where keeping the plant running was the primary focus'.

According to Cullen (cited in *Finding Petroleum* 2013), a failure to translate safety rhetoric into effective action characterised BP Products North America Incorporated's operation at the time of Texas City and Macondo: 'BP [had] the aspirational goal – no accident, no harm to people – but it appears that refinery managers ... had no guidance from corporate-level refinery management as to how to achieve that goal'.

In its report into Macondo, the Deepwater Horizon Study Group (DHSG) distinguished between BP Products North America Incorporated's public announcements about safety and its 'unconscious safety mind':

> Analysis of the available evidence indicates that when given the opportunity to save time and money – and make money – trade-offs were made for the certain thing – production – because there were perceived to be no downsides associated with the uncertain thing – failure caused by the lack of sufficient protection. Thus, as a result of a cascade of deeply flawed failure and signal analysis, decision-making, communication and organisational–managerial processes, safety was compromised to the point that the blowout occurred with catastrophic effects. ... At the time of the Macondo blowout, BP's corporate culture remained one that was embedded in risk-taking and cost-cutting – it was like that in 2005 (Texas City). ... Perhaps there is no clear-cut 'evidence' that someone in BP or in the other organisations in the Macondo well project made a conscious decision to put costs before safety; nevertheless, that misses the point. It is the underlying 'unconscious mind' that governs the actions of an organisation and its personnel. Cultural influences that permeate an organisation and an industry and manifest in actions that can either promote and nurture a high reliability organisation with high-reliability systems, or actions reflective of complacency, excessive risk-taking and a loss of situational awareness. (Deepwater Horizon Study Group 2011: 5–6)

Rousseau (1990) envisioned organisational culture as a series of concentric rings of belief and artifaction. In Rousseau's conception, organisational behaviour is influenced by assumptions, values, norms and expectations. The character of an organisation's output (for example, how a bank teller handles a customer who is overdrawn, or the reliability of a product) reflects its assumptions, values, norms and expectations. With reference to Rousseau's imagining of organisational culture and the conclusions drawn by British Petroleum (2010) and the Deepwater Horizon Study Group (2011), it is clear that neither the culture on board the Deepwater Horizon rig, nor within BP Products North America Inc. was supportive of the mission – the drilling of deep sub-sea wells with reasonable economic efficiency and with a reasonable margin of safety. In July 2016, BP estimated the final cost of the Macondo disaster at $62 billion (Bomey 2016).

TECHNOLOGY

Using fault-tree analysis, BP's internal investigation team identified various technical issues, including the failure of the Blow Out Preventer (BOP): '[T]he BOP emergency functions failed to seal the well after the initial explosions' (British Petroleum 2010: 3). According to BP (2010: 5), the explosion and fire that destroyed Deepwater Horizon originated in a melange of technical and organisational factors: '[A] complex and interlinked series of mechanical failures, human judgments, engineering design, operational implementation and team interfaces came together to allow the initiation and escalation of the accident. Multiple companies, work teams and circumstances were involved over time [including BP, Transocean, Halliburton, Anardarko and Schlumberger/Smith International Inc. (M-I SWACO)]'.

In mitigation, it should be remembered that the extraction of hydrocarbon reserves from the deep ocean carries risk (at the time of the disaster, Deepwater Horizon was drilling in a hurricane-prone region in 5,000 feet (1,500 metres) of water), that every sub-sea drill is unique, that human error is commonplace (Institute of Medicine 2000; Glendon, Clarke and McKenna 2006) and that it is not possible to anticipate every adverse natural or technical circumstance.

BUILDINGS AND INFRASTRUCTURE

As mentioned, BP (2010: 3) identified numerous technical weaknesses and failures, including the failure of the BOP: '[T]he BOP emergency functions failed to seal the well after the initial explosions'. As to why the BOP failed, BP (2010: 5) stated: '[T]he investigation team found indications of potential weaknesses in the testing regime and maintenance management system for the BOP'.

It was unequivocally BP's responsibility to ensure that its contractors worked together harmoniously: 'As the major shareholder, BP was the operator, in charge of overseeing operations of the well. ... Once ... plans were approved and permits ... issued, BP served as the general contractor, in the sense that it was responsible for hiring and overseeing the work of various contractors needed to support the drilling operations and the design and construction of the well. BP personnel were present on board the Deepwater Horizon' (Deepwater Horizon Study Group 2011: 103). Disharmony within BP's contractor workforce (which drew personnel from Transocean, Halliburton, Anardarko and Schlumberger/Smith International Inc. (M-I SWACO)) would have posed a risk to BOP serviceability.

In his speech marking the twenty-fifth anniversary of the Piper Alpha disaster, Lord Cullen, mindful of comments made about contractor co-ordination by the US National Commission on the BP Deepwater Horizon Oil Spill and Offshore Drilling, said: 'BP and other operators needed an effective system in place for integrating the various corporate cultures, internal procedures and decision-making protocols of the many different contractors involved in a deep-water well' (Cullen cited in *Finding Petroleum* 2013). The Macondo disaster suggests an inverse relationship between organisational complexity and operational integrity. Perrow (1984) claims that system behaviour is influenced by interactive complexity. Other things being equal, the greater the complexity of a production system, the greater the likelihood of incident, accident and near-miss.

PROCESSES AND PROCEDURES

As mentioned above, BP (2010) conceded that its procedures for managing sub-contractors and work processes were inadequate. With reference to the work of Professor Jim Reason (1990, 1997, 2013), BP's failure in this regard created a latent error or resident pathogen within the Deepwater Horizon production system – an error activated on 20 April 2010 by adverse geological conditions. The company's internal investigation claimed the disaster resulted from 'a complex and interlinked series of mechanical failures, human judgments, engineering design, operational implementation and *team interfaces. Multiple companies, work teams and circumstances were involved over time* [my emphasis]' (British Petroleum 2010: 31).

CULTURE

According to the Deepwater Horizon Study Group (2010: 9), BP Products North America Inc. failed to cultivate a robust, inclusive safety and learning culture, BP Products having limited its safety horizon to mitigating 'trip-and-fall' risks: '[BP Products North America Inc.'s] system was not propelled toward the goal of maximum safety in all of its manifestations, but was rather geared toward a trip-and-fall compliance mentality rather than being focused on the Big-Picture. ... The system was not reflective of one having well-informed, reporting or just cultures. The system showed little evidence of [having] the willingness and competence to draw the right conclusions from ... safety signals. The Macondo well disaster was an organisational accident whose roots were deeply embedded in gross imbalances between the system's provisions for production and those for protection'.

Grotan (2013: 70) observes of BP Products North America Inc.'s safety horizon: '[S]enior management were not indifferent to safety, but were focused on occupational, rather than process safety'. As Cullen has pointed out, the industry spends too much time on preventing slips, trips and falls and too little time on preventing system or process accidents. This was the case at both Buncefield and Texas City: '*The safety management system [in operation at Buncefield] focused too closely on personal safety* and lacked any real depth on control of major hazards. There should have been an understanding of major accident risk and systems designed to control them. ... The US said [the Texas City disaster] was caused by deficiencies at all levels of the corporation. Cost-cutting, failure to invest and production pressures had impaired process safety performance. ... *The reliance on a low personal injury rate as a safety indicator had failed to provide a true picture of the health of the safety culture* [my emphasis]' (Cullen cited in *Finding Petroleum* 2013).

Dekker (2014a) makes the point that the safest organisations are those whose safety actions are guided by both quantitative indices (such as the number of on-site injuries in a given period) and qualitative indices (such as culture surveys). Regarding BP Products North America Inc.'s performance prior to Macondo, Dekker (2014a: 350–351) notes: '[W]hile measurable safety successes were celebrated, [BP Products North America Inc.'s] coherent understanding of engineering risk across a complex network of contractors had apparently eroded. ... [B]ureaucratic indicators of "safety", particularly quantitative ones, may well suggest that risk is under control – more so than it actually is'. BP Products North America Inc. chose to manage safety through the prism of numbers rather than through the prism of operational research into the *praxis* of production.

Had BP Products North America Inc. possessed a learning culture, it would have implemented the changes recommended in the Baker report into the Texas City disaster. It failed to do so:

> This disaster ... has eerie similarities to the BP Texas City refinery disaster. These similarities include: a) multiple system operator malfunctions during a critical period in operations, b) not following required or accepted operations guidelines ('casual compliance'), c) neglected maintenance, d) instrumentation that either did not work properly or whose data interpretation gave false positives, e) inappropriate assessment and management of operations risks, f) multiple operations conducted at critical times with unanticipated interactions, g) inadequate communications between members of the operations groups, h) unawareness of risks, i) diversion of attention at critical times, j) a culture with incentives that provided increases in productivity without commensurate increases in protection, k) inappropriate cost and corner-cutting, l) lack of appropriate selection and training of personnel, and m) improper management of change. (Deepwater Horizon Study Group 2010: 10)

With reference to the work of Professor Brian Toft (1992, 1997), the best one can say of BP Products North America Inc. is that it achieved passive learning. Following completion of an internal investigation into Texas City (see Mogford (2005)) and publication of the Baker Report (see Baker (2007)), BP Products North America Inc. knew what measures were needed to reduce operational risk. However, it failed to act. It failed to achieve active learning. It failed to demonstrate behaviours associated with organising for high-reliability. It failed to demonstrate mindfulness. It failed.

PEOPLE

Had BP Products North America Inc.'s management implemented the changes recommended in the Baker Report, the Deepwater Horizon disaster might have been avoided. Management failed to take effective action, its narrow conception of safety (see above) leading it to overlook systemic weaknesses. These included:

- poor co-ordination of sub-contractors;
- communication failures (both vertical and horizontal);
- flawed practices;
- an over-reliance on metrics (quantitative indicators of safety performance);
- a culture of risk-taking;
- complicitous regulators.

The Deepwater Horizon Study Group (2010: 10) noted:

> Both [Texas City and Macondo] served … to clearly show there are important differences between worker safety and system safety. One does not assure the other. … [R]isks were not properly assessed in hazardous natural and industrial-governance-management environments. The industrial-governance-management environments unwittingly acted to facilitate progressive degradation and destruction of the barriers provided to prevent the failures. An industrial environment of inappropriate cost and corner-cutting was evident … as was a lack of appropriate and effective governance – by either the industry or the public governmental agencies. As a result, the system's barriers were degraded and destroyed to the point where the natural environmental elements (e.g., high-pressure, flammable fluids and gases) overcame and destroyed the system.

The Deepwater Horizon disaster resulted from human failure at every level.

There are similarities between the systemic origins of Macondo and those of the Piper Alpha and Fukushima disasters. In each case, those responsible for safety failed in their duty to protect employees, the environment and, in the case of Fukushima, the general public. In respect of both Piper and Fukushima, these failings originated at the very heart of government, where a powerful, pervasive and ultimately destructive 'production-at-almost-any-cost' argument emerged from politicians' equating of energy independence with national security.

In the case of Macondo, the same failings originated in various regulatory agencies and the wider industry: 'The organisational causes of [Macondo] are deeply rooted in the histories and cultures of the offshore oil and gas industry and the governance provided by … regulatory agencies. While this particular disaster involves a particular group of organisations, the roots of the disaster transcend this group of organisations. This disaster involves an international industry and its governance' (Deepwater Horizon Study Group 2010: 9).

Civil society played its part by creating demand for hydrocarbons. Americans rely on the automobile. According to the US Census Bureau, in 2013, 86 per cent of Americans commuted to work by car. The automobile has had a profound impact on where and how Americans live: 'The automobile has played a fundamental role in shaping where we live and how we get around. It has influenced the form and density of our communities and expanded the geographic range of daily travel. Nationally, the private automobile is the predominant form of transportation for work and other travel purposes' (McKenzie 2015). If one views the Macondo disaster through a systems-thinking lens, one can see that most Americans were unwittingly complicitous in the disaster. People want the comfort and convenience of personal transport, whether car or motorbike. But no driver or rider wants to pay more than the bare minimum for fuel. The level and nature of demand for hydrocarbons is determined by civil society.

CONCLUSIONS

Like Aberfan, Bhopal, Kings Cross, Challenger, Piper Alpha, Dryden, Fukushima, Texas City, Buncefield and numerous other disasters, Macondo was a system accident. It originated in the behaviour of, and interactions between numerous actants: from car drivers to oil company executives; from the sub-sea blow-out preventer (that manifestly failed) to Deepwater's sub-contractors.

It is interesting to ponder why disasters repeat despite the publication of safety recommendations. Was Perrow correct? Are accidents the norm? Are high-momentum, tightly coupled socio-technical systems incapable of active learning?

Viewing the problem of safety as an insider might see it is illuminating. Even if the will is there to action safety recommendations, the pressure from investors, shareholders, lenders (for example, the banks) and customers to drive down costs is *relentless*. Caught in this maelstrom, managers may find it difficult – if not impossible – to resource safety initiatives. Systems-thinking can help us understand why there is so little active learning.

Macondo was a system accident, rooted in factors such as BP Products North America Incorporated's failure to learn lessons from incidents, accidents and near-misses and failure to scan for systemic weaknesses. Latent errors/resident pathogens included: manning a platform with a fractured workforce; pushing the envelope of technical feasibility; and getting swept along by the industry's laissez-faire, can-do mentality. The company's safety rhetoric (it aspired to zero accidents) created expectations that, in the context of its safety culture and focus on occupational health, could not be met. To the extent that BP Products North America Incorporated was responding to consumer demand for cheap hydrocarbons, Macondo was a *societal* accident. Every consumer of hydrocarbons bore some responsibility for the Deepwater Horizon disaster. Corporations are socially-produced actants. They are shaped by expectation, culture and consumer preference. As Monteiro (2012) puts it: 'You do not go about doing your business in a total vacuum, but rather under the influence of a wide range of ... factors'. There is a symbiotic relationship between consumers and producers.

Safety audits

Safety audits vary in quality. The most effective are those that draw on systems-thinking – that is, the most effective take account of the complexity of socio-technical systems, their failure modes and adaptive/coping mechanisms. The least effective are those that fail to look beneath the surface – that is, the least effective fail to recognise that system performance is influenced by, potentially, n factors, many of which are hidden and/or latent.

In the early 1990s, the University of Texas Human Factors Research Project developed a systems-thinking-informed safety audit methodology for commercial aviation. The Line Operations Safety Audit (LOSA) used trained flight-deck observers to record threats, errors, work-arounds and innovations. The LOSA methodology has proved popular with the aviation industry (International Civil Aviation Organisation 2002; Australian Transport Safety Bureau 2007) and has migrated from the flight-deck to other domains, including:

- air traffic control, where it manifests as the Normal Operations Safety Survey (NOSS) (Henry 2007);
- ramp operations, where it manifests variously as the Dispatch Operations Safety Audit (DOSA), the Ground Operations Safety Audit (GOSA) (Qantas Airways) and the Ramp Operations Safety Audit (ROSA) (Delta Airlines) (Flight Safety Foundation Editorial Staff 2008; Khoshkhoo 2017; Federal Aviation Administration 2018);
- maintenance operations, where it manifests as the Maintenance Line Operations Safety Assessment (M-LOSA) (Continental Airlines) (Ma and Rankin 2012);
- military operations, where it manifests as the Mission Operations Safety Audit (MOSA). The MOSA is a self-assessment tool that does not use observers (Burdekin 2003, 2015).

The International Civil Aviation Organisation (2002: vii) notes: 'In 1999, ICAO endorsed LOSA as the primary tool to develop countermeasures to human error in aviation operations. ... The number of operators joining LOSA has constantly increased since March 2001, and includes major international operators from different parts of the world and diverse cultures'.

Commercial aviation line operations safety audits (LOSAs)

The data required to ensure the safe operation of commercial aircraft is acquired in three ways. First, through attitude surveys. Secondly, through confidential near-miss and error-reporting systems. Thirdly, through line operations safety audits. The Federal Aviation Administration (2018) notes: 'Managing risks has become increasingly important in modern organisations. The aviation industry is maturing in its preference for proactive intervention over post-accident remediation. Systems such as National

Aeronautics and Space Administration Aviation Safety Reporting System (ASRS) and the Maintenance Aviation Safety Action Program (ASAP) encourage air carrier and repair station employees to voluntarily report unsafe conditions. However, those systems are used … following adverse events. Line Operations Safety Assessments (LOSA) address aviation safety proactively'.

A LOSA 'is an observational methodology … which uses expert observers in the cockpit during normal flights to record threats to safety, errors and their management, and behaviours identified as critical in preventing accidents' (Helmreich 2000: 782).

The methodology acknowledges the *ubiquity* of threat and error in aviation. Threats include such things as mechanical breakdown ('tech faults'), adverse weather and sub-optimal air traffic control. Errors include failure to maintain spatial or situational awareness and choosing a course of action that fails to deliver the intended outcome. McDonald, Garrigan and Kanse (2006: 1–2) observe: 'The Threat and Error Management Model … as used in aviation, defines *threats* as *external situations that must be managed by the cockpit crew during normal, everyday flights. Threats increase the operational complexity of the flight and pose a safety risk to the flight at some level.* … The operational definition of flightcrew *error* on the other hand, is *action or inaction that leads to deviation from crew and/or organisational intentions or expectations.* … Threats and errors are considered to be normal parts of everyday operations that must be managed. Threats and errors may sometimes go undetected, on other occasions, they may be effectively managed, or may result in additional errors which require subsequent detection and response'.

The LOSA methodology is an example of a proactive risk management tool. By generating qualitative and quantitative data on operations, LOSA affords managers and front-line personnel the opportunity to address weaknesses (latent errors/resident pathogens) before circumstance sees them manifest as incidents or accidents. For example, a LOSA provides:

- for the identification of overly complex procedures before they cause a crew to lose situational awareness in a critical situation (in 1998 a Swissair

MD-11 caught fire and crashed, killing all 229 passengers and crew. The pilots had become so engrossed in the MD-11's emergency check-lists they lost situational awareness);
- for the identification of mindless air traffic control before it produces a near-miss (an 'air-prox') or worse, and
- for the identification of sub-optimal crew resource management before it leads to a breakdown in communication between the pilots, between the flight-deck and air traffic control or between the flight-deck and cabin.

The International Civil Aviation Organisation (2002: 1/5) observes: 'A medical analogy may be helpful in illustrating the rationale behind LOSA. ... Periodic monitoring of routine flights is ... like an annual physical: proactively checking health status in an attempt to avoid getting sick. Periodic monitoring of routine flights indirectly involves measurement of all aspects of the system, allowing identification of areas of strength and areas of potential risk'.

To summarise, a LOSA is a systems-thinking-informed, proactive, inclusive and organic intra-organisational safety tool that records threats, errors, work-arounds (expedients) and *in-situ* innovation. One of the unique features of a LOSA is that it records good practice, unlike traditional quality-control methodologies. That is, a LOSA records both the things flight and cabin crew do badly and the things they do well. It records work-arounds and innovations. The LOSA methodology's cycle of observation, analysis, critique, learning and application helps participating airlines improve their safety performance.

The LOSA methodology draws on theories of:

- reflective practice (see, for example, Dewey (1933), Schön (1973, 1983) and Bolton (2010));
- organisational mindfulness (see, for example, Weick et al. (1999) and Weick and Sutcliffe (2007));
- communities of practice (see, for example, Wenger (1998)), and
- active learning (see, for example, Toft (1992) and Toft and Reynolds (1997)).

The International Civil Aviation Organisation (2002: vii) notes: 'A particular strength of LOSA is that it identifies examples of superior performance that can be reinforced and used as models for training. In this way, *training interventions can be reshaped* and reinforced *based on* successful performance, that is to say, *positive feedback*. This is indeed a first in aviation, since the industry has traditionally collected information on failed human performance, such as in accidents and incidents [my emphasis]'.

The LOSA methodology provides for the creation of a safety regime grounded in both negative *and* positive feedback. That is, on understanding why things go wrong and why things go right. The methodology embodies Hollnagel's Safety-II model. It mines good practice and bad practice for lessons. Its field of view is 360°. Table 2 describes how a LOSA is performed.

Table 2. How a LOSA is performed.

LOSA Stages		
1	Resourcing	Resources of time and money are appropriated for the purpose of performing a LOSA. Audits are costly. To secure flight-crew buy-in the LOSA is overseen by a steering committee composed of managers *and* pilots.
2	Observation	Disinterested, trained observers record threats, errors, adaptations and innovations on scheduled flights. The observers are often the airline's own pilots. Observations are made with the consent of the flight crew. Observers complete a narrative account and code threats and errors. Records are de-identified. No personal data is recorded.
3	Analysis	Returns are analysed by an internal committee of experts that includes pilots. Outliers are discussed and trends identified. Data is cleansed (to assure consistency).
4	Critique	The data is used to evaluate the airline's operational orders, systems and performance.
5	Learning	Orders and systems are modified or abandoned, or new orders and systems introduced, as appropriate.
6	Application	Pilots are trained in the new orders and systems prior to them being actioned.
7	Resourcing	See 1, above.
8	Observation	See 2, above. The six stages represent a virtuous circle of problem identification and remediation (see Figure 19).

Systems-thinking in practice

Figure 19. The Line Operations Safety Audit virtuous circle, from resourcing the audit to organisational change.

No safety-assurance methodology is problem-free. The LOSA is no exception. The major issue is the possibility of a Hawthorne effect – observees modifying their behaviour and/or the way in which the flight is managed (for example, scheduling workload such that the more experienced pilot flies the more difficult sectors) (Todd and Thomas 2012). There are mitigations. Thoughtful rostering, where observations are made by pilots unknown to observees, helps reduce the risk of a Hawthorne effect. Other risks inherent in a LOSA include:

- the skewing of the data by experimenter bias (bias towards the observer): observees' identification with the observer could cause them to modify their behaviour;

- the skewing of the data by observer bias (bias towards observees): observers' preconceptions and/or personal loyalties may influence how an episode is interpreted;
- the skewing of the data by the LOSA coding structure: to a degree, coding systems circumscribe researchers' field of view;
- the skewing of the data due to cognitive overload. Flight-decks are busy workspaces. There is much to take in. Overload could cause the observer to misinterpret an episode or fail to capture data. Poor prioritisation and information overload cause task saturation, reducing situation awareness.

Railway safety audits

The LOSA methodology (Klinect, Wilhelm and Helmreich 1999; International Civil Aviation Organisation 2002) has been adapted for use in various industries, including the rail industry. In a 2006 conference paper titled *Confidential Observations of Rail Safety (CORS): An Adaptation of Line Operations Safety Audit*, McDonald, Garrigan and Kanse (2006) describe an evaluative study conducted in collaboration with Australia's Queensland Rail (Queensland Rail celebrated 150 years of passenger service in 2015).

Developed with input from the University of Queensland and the Australian railway industry, the Confidential Observations of Rail Safety (CORS) methodology is informed by Klinect, Wilhelm and Helmreich's (1999) and the ICAO's (2002) LOSA methodology, specifically the Ten Operating Characteristics (TOCs). The TOCs are described in Table 3.

Table 3. LOSA's TOCs (McDonald, Garrigan and Kanse 2006: 2–3; the table is reproduced exactly as drafted by the authors).

	LOSA's Ten Operating Characteristics	CORS
1	Jump-seat observations during normal flight operations	CORS observations are conducted on normal timetabled revenue services, with participants fulfilling their normal rostered workings. There are no observations conducted during driver monitoring or assessment journeys.

	LOSA's Ten Operating Characteristics	CORS
2	Joint management/pilot sponsorship	A Steering Committee was developed during the planning stage of CORS. This provided an opportunity for management and union involvement and endorsement of the program. The Steering Committee is involved in all stages of CORS, including development of observation materials, observer selection and training, scheduling of observations, review of progress, and interpretation and communication of results.
3	Voluntary crew participation	CORS observations are only conducted with permission of the participating driver. When observers approach drivers, they are required to provide information about CORS, and obtain voluntary informed consent prior to conducting an observation. If a driver declines, the observer continues to approach other drivers until consent is obtained. Participants may withdraw from the program at any stage during an observation.
4	De-identified, confidential and safety-minded data collection	CORS observation worksheets were designed to ensure that no identifying information is recorded, such as names, train numbers or dates of observation.
5	Targeted observation instrument	CORS observers use specially designed observation materials based upon the Threat and Error Management Model, and target external threats, crew errors, how these are detected and managed, and actual outcomes. Observers record and code events according to threat and error code lists specifically developed for this project by driver representatives and Human Factors experts.
6	Trusted, trained and calibrated observers	To promote trust in CORS, observers are peer train drivers, selected from expressions of interest, and specially trained in the technical and non-technical skills required for conducting observations. Observers participate in regular calibration sessions to promote consistency and reliability of coding.

(*Continued*)

Table 3. (*Continued*)

	LOSA's Ten Operating Characteristics	CORS
7	Trusted data-collection site	All observation worksheets and signed informed consent forms are sent to the University of Queensland for data entry and storage. No one within the rail organisation has access to individual observation paperwork. Feedback from drivers has shown that this discernable level of independence greatly contributed to their confidence in, and acceptance of the CORS program.
8	Data verification roundtables	Calibration sessions are conducted regularly to promote reliability and consistency of coding. These sessions have also provided observers with an opportunity to discuss techniques that worked well when approaching participants, or conducting observations, and to explore any barriers or difficulties experienced.
9	Data-derived targets for enhancement	At the time of writing the paper, CORS observations were not yet complete, however, the next stage of CORS involves analysing the data obtained in the first round of observations to determine trends or key issues, and to prioritise these for attention. An action-plan will be developed to address key issues that have emerged.
10	Feedback of results to the line pilots	Results of the data analysis will be communicated to drivers, and other key stakeholders within the organisation via the Steering Committee. It will be important to link these results to planned improvement actions to demonstrate that the program has been used to generate change and safety improvement.

Regarding the pioneering work done by the University of Queensland with Queensland Rail, one of the most interesting findings was that the threat-and-error profile of rail operations is broadly similar to that of air operations. A LOSA survey of three airlines noted a threat-rate per sector of 3.3, 2.5 and 0.4, and an error-rate per sector of 0.86, 1.9 and 2.5 (Klinect, Wilhelm and Helmreich 1999). The CORS survey of Queensland Rail noted a threat-rate per journey of 4.6, and an error-rate per journey of 1.6 (McDonald, Garrigan and Kanse 2006: 5).

According to McDonald, Garrigan and Kanse (2006: 5), the CORS methodology shows promise: '[T]he [CORS] programme appears to be well-accepted among all stakeholder groups, observers have no difficulties in recruiting participants, and the collected data is starting to provide insights not only into which threats and errors occur, but also how well these are managed'.

While systems-thinking-informed safety-assurance methodologies such as LOSA and CORS show promise, it should be noted that they require significant investments of time and money. Training observers is costly. Calibrating observers is costly. Storing and analysing data is costly. Remedying identified weaknesses is costly. If the necessary resources for a LOSA or a CORS cannot be secured, it is better that the audit is postponed. A LOSA or a CORS implemented on the cheap, or executed in a half-hearted manner, could do more harm than good, especially if there are no funds for remediations. These methodologies require *significant* investments of enthusiasm, time and money to produce results.

Following a series of high-profile air disasters in the 1960s and 1970s, regulators and airlines decided to act. Responses included the team-working standard crew resource management (CRM) and a proactive risk-management tool, the line operations safety audit (LOSA). The line operations safety audit's proactivity provides for the elimination of latent errors/resident pathogens before they morph into near-misses, incidents or accidents. Because a LOSA identifies good as well as bad practice, it helps airlines become more efficient. Good practice can be documented, worked into procedures, trained and rolled-out across the airline. A LOSA requires the prior, informed consent of observees who must be assured that data will be de-identified and not used punitively. The *sine qua non* of a LOSA is trust. The project will stall if observees are distrustful. The methodology has been successfully migrated to other industries.

CHAPTER 3

A case study in systems-thinking

Introduction

In the same way that technologies reflect our values and priorities (a motorway scything through the countryside reflects society's prioritisation of economic growth over conservation), near-misses, incidents and accidents are a product of the social, economic, political and techno-scientific milieu (see Figure 20).

Figure 20. Adverse events have complex origins (after Challenger, Clegg and Robinson (2009)).

A systems-thinking-informed analysis of the 2017 Grenfell Tower, London, fire disaster

Introduction

In the small hours of 14 June 2017, a fire spread rapidly through a London tower block, killing seventy-one people (see Figure 21). On 14 December 2017, a memorial service to honour the victims of the Grenfell Tower fire disaster was held at London's St Paul's Cathedral, a venue synonymous with public expressions of grief and thanksgiving. Attendees included victims' relatives and friends, the Prince of Wales, the Duke and Duchess of Cambridge, the Prime Minister and the Leader of Her Majesty's Opposition.

Figure 21. An obscenity in the world's fifth largest economy? (Wikimedia Commons 2018l).

The daughter of a 78-year-old resident of Grenfell Tower who lost her husband in the fire said: 'It's very, very hard. Still [my mother] cries every day. ... It's six months and it's still very hard for us' (Jafari cited in Agerholm 2017). Absent from the service was the Leader of the Royal Borough of Kensington and Chelsea (RBKC) Council, the local authority responsible for maintaining Grenfell Tower. Council Leader Elizabeth Campbell stayed away at the request of victims' families.

By the end of 2017, the disaster, already subject to investigation by the Metropolitan Police, became the focus of a Public Inquiry. Led by retired judge Sir Martin Moore-Bick, the Inquiry aimed to publish an Interim Report by the Autumn of 2018. In a 10 August 2017 letter to the Prime Minister, Moore-Bick addressed the question of the Public Inquiry's terms of reference (ToRs). Moore-Bick (2017: 2) observed: '[M]any of those who have been affected by the fire, and some others, feel strongly that the scope of the Inquiry should be very broad, and should include an examination of social housing policy and all aspects of the relationship between the residents of the Lancaster West estate on the one hand, and the tenant management organisation on the other'. Despite claiming to 'well understand why local people consider that these are important questions which require urgent examination' (Moore-Bick 2017: 2), Moore-Bick rejected calls for a broad investigation.

Moore-Bick's decision, supported by the Prime Minister, drew hostile comment. Kensington's Member of Parliament accused the government of commissioning a 'technical assessment' that would fail to get to the 'heart of the problem' (Dent-Coad cited in Pasha-Robinson 2017). Tottenham's Member of Parliament said: 'I am deeply disappointed that the narrow terms of reference ignore the issue of the provision and management of social housing in the UK. The decision to duck these important issues and fail to learn the broader lessons of Grenfell represents a grave injustice to those who died' (Lammy cited in Pasha-Robinson 2017). According to Pasha-Robinson (2017), the Prime Minister was prepared to consider using other mechanisms to investigate causal factors.

For Professor Emeritus Colin Crouch, Grenfell symbolised the failure of the politics of neoliberalism: 'For many people in, or around Britain, the sight of the burning hulk of the Grenfell Tower block of flats ... was the final horrific comment on the ideology that had guided so much public policy for the previous four decades' (Crouch 2017: 1). According to Crouch, the Grenfell Tower fire was a 'self-made disaster' – the result of decades of deregulation, low taxes, cost-cutting, running-down of the public sector and social and economic polarisation. Britain, claimed Crouch (2017: 3), had become a 'low-tax, low-regulation regime', harbouring 'high inequality' and exhibiting a 'lack of concern for collective needs'. According to the Trust for London, London is Britain's most economically polarised city. According to the Social Integration Foundation, London is Britain's most socially polarised city (*The Observer* 2015). In the nineteenth century, Charles Dickens documented London's social and economic polarisation. *Plus ça change*?

Others shared Crouch's opinions. Chris Williamson MP, shadow minister for the emergency services, observed:

> [I]f ... we intend to leave no stone unturned in understanding what 'went wrong' then let's not shy away from examining the underlying principles that have guided government for the past 30 years. In the mid-1980s, Thatcher's government deregulated fire safety standards in homes, abandoning enforceable requirements for guidelines which the building industry could choose to follow, or not. Also complicit was the government led by Tony Blair, who, ideologically speaking, was the offspring of his Tory predecessors. Despite New Labour's gains for working people ... behind the scenes the same principle of deregulation was at work. Blair's government further blurred the lines of responsibility for fire safety enforcement by insisting that the role of a fire inspector should be to 'inform and educate' rather than enforce. With a renewed emphasis on facilitating the private sector, the role of fire safety enforcement officers morphed into being that of 'business support'. ... [D]uring the mid-80s ... the imperatives of private profit began to trump the need for public safety. (Williamson 2017)

Williamson, like the Leader of Her Majesty's Opposition, the Mayor of London and others, wanted the Moore-Bick Public Inquiry to consider any factor pertinent to the disaster – including the impact on public safety of Britain's neoliberal political consensus. The Labour Party, in a nod to systems theory, called for two inquiries: one to investigate the disaster's immediate causes; another to investigate the disaster's proximate causes (Williamson 2017).

The government did two things. First, it instructed Dame Judith Hackitt, Chair of the Engineering Employers' Federation and Fellow of the Royal Academy of Engineering, to review the building regulations (Hackitt 2017, 2018). Hackitt, on being appointed, said: 'I look forward to working with *experts* from across different sectors to take an urgent, fresh and comprehensive examination of the regulatory system and related compliance and enforcement issues [my emphasis]' (Hackitt cited in Pitcher 2017). Secondly, it instructed Sir Martin Moore-Bick, a retired judge, to conduct a Public Inquiry.

Both investigations were circumscribed. The Hackitt review made no provision for mining the experiences of those who survived the Grenfell Tower fire (despite representations from legal counsel) (see Figure 22). The Moore-Bick inquiry made no provision for investigating the proximate causes of the Grenfell Tower disaster (Moore-Bick 2017). *The Sun's* Amanda Devlin (2018) noted: 'Survivors believed [Moore-Bick] should investigate the social and political conditions which allegedly allowed Grenfell to happen, but Moore-Bick said such issues were "not suitable for a judge-led inquiry"'. The chair of the Royal Institute of British Architects (RIBA) observed: 'It is disappointing … that the terms of the [Moore-Bick] Inquiry do not explicitly mention the overall regulatory and procurement context for the construction of buildings in the UK. We consider this examination crucial to understanding the often complicated division of design responsibilities and the *limited level of independent oversight of construction*. These pervade many current building procurement approaches prevalent in the public and housing association sectors. Such *regulatory and procurement concerns should not be dismissed* as they would have helped set the full context for the decisions that were made at Grenfell Tower and at other residential buildings. This is the missing piece of the puzzle [my emphasis]' (Duncan 2017).

A systems-thinking-informed investigation would have looked very different to the Moore-Bick public inquiry. For example, it would have considered the possibility that agencies responsible for regulating the building industry had been captured by those subject to regulation (regulatory capture). Further, it would have investigated the social, economic and political forces that – intentionally or unintentionally – ghettoise London's residents.

Figure 22. The charred hulk of Grenfell Tower photographed nearly eleven months after the fire (Wikimedia Commons 2018m).

Dame Judith Hackitt's review, published in May 2018, was criticised for not recommending an outright ban on combustible building materials. The Chair of the House of Commons Housing, Communities and Local Government (HCLG) Committee said: 'While the Independent Review has come to many sensible conclusions, I strongly regret Dame Judith's decision not to recommend an immediate ban on the use of combustible materials in the cladding of high-rise residential buildings. The approach she proposes places too much faith in the professional competence of a construction industry in which too many people have been inclined to take shortcuts and cut costs at the expense of the safety of residents' (Betts cited in Lawrence 2018). Was the Hackitt review a victim of regulatory capture?

The systems-thinking-informed analysis presented below is informed by Challenger, Clegg and Robinson's (2009: 90) conception of the network space as comprising goals, people, buildings/infrastructure, technology, culture and processes/procedures.

Systems-thinking-informed analysis

GOALS

According to Tucker (2017), 'Grenfell was a … political tragedy'. The Grenfell Tower disaster occurred against a backdrop of changing values and priorities *vis-à-vis* social capital, council tenants and public safety (Crouch 2017; Gapper 2017; Horton 2017; Hughes 2017; McKee 2017; Möller 2017; Tucker 2017; Williamson 2017; Hackitt 2017, 2018). Specifically:

- the drive to create a property-owning democracy. In 1951, the Conservative Party declared its intention to create a property-owning democracy. Some three decades later, Mrs Thatcher actioned the policy by offering tenants the right to buy their council-owned house or flat. From that moment, council-owned housing and council tenants were framed as a problem. For most of the post-war period, social housing had been considered part of the solution to Britain's housing crisis. From the early

1980s onwards, governments of the right and the left framed council-owned housing as a social drag. Council houses and council tenants were problematised and stigmatised. Governments set about reducing the volume of council-owned properties. Houses and flats were sold at a discount. The rate at which new council houses were built fell. The reduction of council provision, in the context of growing inequality, produced ghettoisation. One of the consequences of the social and economic polarisation of the last forty years has been the colonisation of council estates by the economically marginalised. While no local authority intends this, it is an inevitable consequence of the widening economic gap between the middle class and the working class (Crouch 2017) and of councils' legal duty to house the poor, homeless and vulnerable: 'A shortage of public housing, alongside the duty of local councils to house the homeless, poor and vulnerable, leads to what is known as "residualisation" – council homes are increasingly allocated to people on very low incomes, rather than the broader spread of the past' (Gapper 2017). Horton (2017) refers to London's working class as being 'packaged and parked' in council-owned housing: 'Those living in Grenfell Tower were a diverse mix of Moroccans, Eritreans, and more, conveniently packaged and parked with few protections by a council that revelled in its exclusive status'. Ghettoisation is a betrayal of Nye Bevan's utopian vision for social housing: 'Council housing was conceived and built in the post-war era as a mainstream way of living – a place "where the doctor, the grocer, the butcher and farm labourer all lived on the same street", in the words of Nye Bevan, the politician most closely associated with the state-managed housing revolution. Primarily for the working class, it also housed many of the more comfortably-off as well, and was not a derided way of living. Indeed, in 1979 one in five of the richest tenth of the UK's population lived in social housing' (Tucker 2017). Ghettoisation was a feature of Britain's nineteenth-century social landscape. Disraeli, in his 1845 novel *Sybil*, wrote of: 'Two nations between whom there is no intercourse and no sympathy; who are ignorant of each other's habits, thoughts and feelings, as if they were dwellers in different zones or inhabitants of different planets; who are formed by different breeding, are fed by different food, are ordered by different manners, and are not governed by the same laws'. If Gapper (2017) and Horton (2017) are correct, Britain's city-dwellers are as polarised today as they were in the time of Disraeli, Dickens and Engels. Perhaps more so, given that tower blocks have no thoroughfares;

- the drive to cut so-called red-tape. The drive to cut red-tape was seen by some as an opportunity to dilute safety legislation. It is possible that politicians' framing of legal protections as irritants rather than panaceas has encouraged partial or non-compliance. There has been a move away from prescription. For example, Part B4, Schedule 1 of the 2010 Building Regulations states that: 'The external walls of the building shall adequately resist the spread of fire over the walls and from one building to another, having regard to the height, use and position of the building. ... The roof of the building shall adequately resist the spread of fire over the roof and from one building to another, having regard to the use and position of the building'. One can only wonder what the term 'adequately resist' means in practice? Regarding performance assessment of building components and materials, desk-top studies have substituted for testing. 'The term "desktop study" ... describe[s] an assessment in lieu of test' observes Hackitt (2018: 93). Regarding the integrity of desktop studies of a material's performance under heat stress, the technical director at the UK Fire Protection Association said: 'I have heard expressions of concern about some of these assessments, for some time, that they may not be as rigorous as they ought to be' (Glockling cited in Tubb 2017). Safety roles have been contracted-out: 'Deregulation ... meant outsourcing responsibility for fire inspections to owners and builders, instead of civil servants. Private companies were required to use "authorised inspectors" for fire safety' (Erlanger 2017). Contracting-out the fire inspection role to owners and builders created a conflict of interest. Potentially, risks to public safety could be downplayed or ignored. Lord Cullen, during his investigation into the 1989 Piper Alpha disaster, drew attention to a potential conflict of interest in respect of the safety oversight of North Sea installations. At the time of the disaster, the oversight role was vested with the department of state responsible for cultivating the industry: '[T]he fact that the safety inspectors were part of the same department promoting the maximisation of production provided at least the potential for an unacceptable conflict of interests. The inspectors, from the inception of the Petroleum Engineering Division of the Department of Energy in 1977, are seen as having worked with the industry in concentrating on disaster prevention to the exclusion of occupational safety so that security of production was guaranteed. To this end the inspectorate rejected the option of prosecution for breach of regulations in favour of a more "softly-softly" approach' (Paterson 1997: 23). Apparently, no one in government saw fit to apply the lessons from the Piper Alpha disaster – so comprehensively documented in Lord Cullen's report – to the construction industry. Whither joined-up government?;

- the hollowing-out of local government. Various local government functions, including those pertaining to the expansion and management of the housing stock, have been contracted out: '[C]hanges in local government [have] included marketisation, a stronger affirmation of risk-taking, and new modes of governance including contractualisation (the outsourcing of core local government functions to corporations or Arms-Length Management Organizations and other Special Purpose Vehicles). UK local authorities have also embraced privatised urban regeneration and the use of highly contested Private Finance Initiatives' (Möller 2017). The hollowing-out of local government has been part of a larger cross-party effort to roll back the state. The deputy director of the Centre for European Reform speaks of a '40-year drive to marginalise or discredit the state and its role in the economy and society' (Tilford cited in Erlanger 2017);

- the exclusion of the consumer voice from decisions pertaining to the refurbishment and redevelopment of council-owned housing estates. McKee (2017) notes: 'It is the powerful who define ... what policies are acceptable, and even whose lives are important. They set the rules that relax standards on safety and employment rights. And they silence the weak, ignoring or discounting their views'. According to a member of campaign group Justice 4 Grenfell, those responsible for managing Grenfell Tower, including the Kensington and Chelsea Tenant Management Organisation (KCTMO), largely ignored residents' views: 'It was not that we stayed silent, but that they never responded. It was

not just that they ignored us, but that they viewed us with contempt' (Williams cited in Gapper 2017). Regarding consultation around housing development projects, Hackitt (2018: 5) noted: 'When concerns are raised ... by residents, they are often ignored';
- reducing public expenditure. In her speech to the 2018 Conservative Party conference, Prime Minister Theresa May promised an end to austerity: 'The British people need to know that the end is in sight. ... Because you made sacrifices, there are better days ahead. ... A decade after the financial crash people need to know that the austerity it led to is over and that their hard work has paid off' (May cited in Youle, Gray, Slawson and White 2018). At the time of the Grenfell disaster, austerity was a dominant theme of right-wing politics. Möller (2017) talks of 'the persistence of austerity as a dominant policy paradigm'. The public services were a prime target for cuts. Housing suffered. Möller (2017) refers to the 'systemic neglect of social housing investments in Britain'. While the Royal Borough of Kensington and Chelsea had spent money on improving the thermal efficiency of Grenfell Tower, it had not invested in new safety measures such as a sprinkler system. Around £10 million was spent on a new heating system, double glazing and exterior thermal cladding. A sprinkler system for Grenfell would have cost a couple of hundred thousand pounds: 'Experts have suggested that sprinklers could have been fitted in the 24-storey building for £200,000 during the £10 million refurbishment' (Hughes 2017). According to Möller (2017), in 2017 the Royal Borough of Kensington and Chelsea had a financial reserve of £274 million. Lagadec (1982: 495) observes: '[T]he disaster, most often, is anticipated, and on multiple occasions'. The Grenfell Tower disaster was foretold by the 2009 Lakanal House fire in Camberwell, south London, that killed six residents. As at Grenfell Tower, the fire spread quickly through the 1958-built high-rise, and in unpredictable ways: '[The fire] spread unexpectedly quickly, both laterally and vertically, combusting the cladding and travelling between apartments' (Gorse and Sturges 2017: 73). The Lakanal inquest coroner, judge Frances Kirkham, suggested to the communities secretary, Eric Pickles, that landlords should consider retrofitting residential blocks with sprinkler systems (Walker 2013);

- remodelling urban spaces. For much of the post-war period, the various agencies and professions concerned with housing (local authorities, new town development corporations, town planners and architects) were in thrall to modernism. Terraced streets and corner-shops were razed. Steel-framed, system-built council houses, tower blocks and shopping malls were erected. The possibility of rehabilitating Victorian or Edwardian properties was seldom considered (much to the annoyance of Sir John Betjeman and his regeneration acolytes). Britain's postwar settlement – a consensus on the country's social, economic and political direction of travel – drew on modernist ideas, both in regard to the production of goods and services and housing (Hall and Jacques 1989). According to Murray (1989: 41), the post-war period was characterised by 'destructiveness in the name of progress'. Britain was gripped by an 'exploitative environmental settlement' that was 'embedded in the industrialism of the big factory and the overpowering modernism of the tower block' (Hall and Jacques 1989: 27). Britain's first tower block was erected in Harlow, Essex, in 1951. Despite the failure of the Ronan Point tower block in east London in 1968 (a gas explosion caused the progressive collapse of one corner of the tower, killing four), local authorities continued to build high. Grenfell Tower was completed in 1974 as part of the Lancaster West slum-clearance project in North Kensington.

PEOPLE

Dame Judith Hackitt, while criticised for not recommending the banning of combustible materials from refurbishment projects, was praised for drawing attention to the human, organisational, systemic and cultural failings that contributed to the Grenfell Tower disaster. Specifically, Dame Judith drew attention to:

- project managers' prioritising of value and speed over quality and consultation (with tenant groups): '[T]he primary motivation is to do things as quickly and cheaply as possible rather than to deliver quality homes which are safe for people to live in. When concerns are raised, by others involved in building work or by residents, they are often ignored.

Some of those undertaking building work fail to prioritise safety, using the ambiguity of regulations and guidance to game the system [my emphasis]' (Hackitt 2018: 5);
- the regulator's reluctance to implement meaningful sanctions against contractors who transgress building regulations: 'Where enforcement is necessary, it is often not pursued. Where it is pursued, the penalties are so small as to be an ineffective deterrent' (Hackitt 2018: 5);
- a general lack of integrity and application on the part of those involved in development projects: '[R]egulations and guidance are not always read by those who need to, and when they do, the guidance is misunderstood and misinterpreted' (Hackitt 2018: 5);
- the negative consequences of a heavily sub-contracted and fragmented work process: '[T]here is ambiguity over where responsibility lies, exacerbated by a level of fragmentation within the industry, and precluding robust ownership of accountability' (Hackitt 2018: 5). Some claim that the sub-contracted, fragmented work regime aboard the Deepwater Horizon platform sowed the seeds of disaster.

The Grenfell Tower disaster occurred against a backdrop of contracting-out. The Royal Borough of Kensington and Chelsea had contracted-out management of Grenfell Tower to the not-for-profit Kensington and Chelsea Tenant Management Organisation. The KCTMO had been set up in 1996 to manage 9,500 properties. According to Gapper (2017), the KCTMO 'had a fractious relationship with those living in Lancaster West'. The KCTMO selected Rydon, a contractor specialising in social housing, to oversee the thermal-efficiency refurbishment of Grenfell Tower. According to Gapper (2017), Rydon got the contract by undercutting rival Leadbitter. Rydon sub-contracted elements of the project to specialist firms: '[A] web of consultants and subcontractors took on parts of the project. … Studio E … were the architects, a task that included specifying materials. The French design firm Artelia acted as employer's agent, representing the Kensington and Chelsea Tenant Management Organisation, and quantity surveyor. Rydon handed the £2.5m job of installing the new façade to Harley Curtain Wall … which was taken over by a linked group, Harley Facades, during the Grenfell refurbishment. CEP Architectural Facades, a division of the

UK manufacturer Omnis, cut the cladding, which was supplied to Harley by the US aluminium group Arconic; insulation was supplied by Celotex, based in Suffolk' (Gapper 2017).

The RBKC created an actor-network for the purpose of managing and maintaining Grenfell Tower. In 2014, a key actant in that network – the KCTMO – created a bespoke actor-network to prosecute the refurbishment. One of the actants in this bespoke network, Rydon, created another actor-network to see the project through. Rydon's corporate actants included Studio E, Artelia, Harley Curtain Wall, CEP Architectural Facades, Arconic and Celotex. According to Gavin Smart, deputy chief executive of the Chartered Institute of Housing, complexity has become a feature of development projects: 'On any major project, public or private, there is an ecosystem of contractors and subcontractors based on their skills and expertise. It is always a complex exercise' (Smart cited in Gapper 2017).

In the aftermath of the disaster, local councillors were accused of dismissing residents' concerns about fire safety at Grenfell Tower (Bennhold 2017). Whether true or not, there is no doubting that councillors presided over a socially, economically and politically divided borough. At the southern end of the Royal Borough of Kensington and Chelsea lies some of the most expensive real-estate in the world. The contrast with dilapidated north Kensington could not be starker: 'Kensington and Chelsea is a microcosm of a divided Britain. The south is home to Kensington Palace Gardens, better known as Billionaires' Row, one of the most expensive streets in the country. Roman Abramovich, the Russian billionaire, owns a mansion there reportedly worth £125 million ($163 million). And Kensington Palace is where Prince William and the Duchess of Cambridge will be raising their children. To the north, Golborne ranks as one of the two poorest wards in London. Victorian-era diseases like tuberculosis and rickets have made a comeback. Life expectancy in parts of North Kensington is 20 years lower than in South Kensington' (Bennhold 2017).

In 2017, thirty-seven of the council's fifty members were Conservatives. Forty-six councillors were white. The council's cabinet was entirely white. After the disaster, local labour councillors accused the Tory group of being out of touch. One observed: 'They don't know how the other side lives' (Press cited in Bennhold 2017). The council's new Tory leader admitted that during her eleven-year career in local government she had never set foot in any of the borough's high-rises. It is worth asking whether those who decided how and where money was spent in the Royal Borough were sufficiently in touch with the needs of its poorer residents. Were they capable of empathising with them?

BUILDINGS/INFRASTRUCTURE

Like many towers, Grenfell's flats were arranged around a central core. The core contained two lift shafts and a staircase (Bulman 2017). Grenfell Tower harboured several latent errors/resident pathogens. For example:

- a design that lacked redundancy. In the field of engineering design, redundancy, whereby components essential to the functioning of a device are duplicated or triplicated, reduces the risk of catastrophic failure. In complex systems such as aircraft, the principle is applied to both human and non-human components. Thus, all commercial aircraft with more than 19 seats must have two pilots. Because Grenfell had a single staircase, there was no redundancy in regard to foot access through the building's twenty-four floors. Grenfell's design complied with planning laws. Safety regulations permit single-staircase towers;
- there was no building-wide fire alarm system;
- there was no sprinkler system. The fire started on the fourth floor in flat 16. Mendick (2017) notes that 'Experts [believe] sprinklers inside could have stopped the initial fire'. Gorse and Sturges (2017: 75) note that 'active fire protection in the form of sprinkler systems are more effective in checking fire-spread than the attendance of numerous fire engines';
- the installation of thermal cladding over the tower's fourteen exterior columns created a chimney, allowing smoke and flame to quickly spread up the sides of the building: 'A total of ten columns run up the sides of the building, with a further column at each corner. Investigators ... believe that when those columns were fitted with cladding, they could have created an air gap that acted as a chimney. Videos of the blaze clearly show the flames spreading upwards far faster than they spread

sideways' (Mendick 2017). The emergency services received their first call at 00:54. Firefighters arrived on-scene in around six minutes to witness the fire spreading to the cladding (Bulman 2017);

- thermal cladding that was combustible. The aluminium cladding's polyethylene filling was combustible (Gorse and Sturges 2017). At the Moore-Bick public inquiry, a lawyer observed of Arconic's product: 'Our understanding is that the ignition of the polyethylene within the cladding panel produces a flaming reaction more quickly than dropping a match into a barrel of petrol' (Barwise cited in Osborne and Agerholm 2018). Hanging combustible panels on the building's exterior concrete walls increased the building's risk-loading;
- fire doors that were non-compliant. Over 100 of the tower's fire doors were non-compliant with the building regulations. According to Booth, Bowcott and Davies (2018) 'Many of them failed in 20 minutes instead of the 60 minutes required by regulations'. Gorse and Sturges (2017: 74) observe: 'The most widely used principle of passive fire protection is that of compartmentation. ... This strategy involves fitting fire doors to all rooms with the necessary fire endurance, usually at least thirty minutes'.

TECHNOLOGY

There are qualitative differences between Victorian and Edwardian terraced properties and high-rise blocks in terms of how they respond to fire and explosion. As a system of residential accommodation, terraced properties are less interactively complex and less tightly coupled than high-rise blocks, where dozens of flats (sometimes of varying configurations) are stacked one upon the other. Perrow (1984) posits an inverse relationship between coupling/complexity and reliability/resilience. That is, the more complex and tightly coupled a system, the less reliable and resilient it will be.

A gas explosion in a terraced house might demolish one or two adjacent properties. A gas explosion in a high-rise might demolish an entire wing. This is what happened at Ronan Point, where the entire corner of a twenty-two-storey block collapsed, killing four and injuring seventeen. The manner in which the block failed – progressive collapse – was a product of interactive complexity, tight coupling and gravity. At the time, the failure was blamed on poor design and careless construction. On 31 December 2018, a gas explosion in a ten-storey tower block in the Russian city of Magnitogorsk destroyed forty-eight flats. In a failure mode identical to that observed at Ronan Point in 1968, the forty-eight flats concertinaed to the ground, killing and trapping sleeping tenants: 'The blast, at 06:02 local time … is thought to have ripped through the first floor, which houses some offices, and the seven storeys above then collapsed' (British Broadcasting Corporation 2018a).

A fire in a terraced house can usually be prevented from spreading (provided the emergency services arrive in good time and are effective). A fire in a high-rise has the potential to destroy multiple flats, especially if, as at Grenfell Tower, it starts on a lower floor. Coupling increases a fire's destructive potential. (Grenfell's under-performing fire doors increased its vulnerability).

Because of coupling and complexity, high-rise blocks carry more risk than terraced housing. At Grenfell Tower, this risk was augmented by a chimney-effect: the aluminium cladding was installed in such a way that it facilitated the spread of smoke and flame up the exterior of the building. Grenfell's cladding tightened the building's coupling.

Given the embedded ('hard-wired') vulnerabilities of tower blocks, it is reasonable to ask why we build them. Because of coupling, complexity, density and the difficulties of firefighting high above the ground, tower blocks carry greater risk than traditional terraced, semi-detached or detached houses. Modularity boosts resilience. Integration diminishes it: '[Modularity anticipates] the inevitable failure of some parts of a system e.g. because of design mistakes, operator errors, faulty parts or supplies, poor procedures, or an unfriendly environment. It … [decomposes] … vertical integration into modules, such that failures in one module [do] not cascade through the system. … [Modularity loosens] the coupling of parts' (Perrow 2012).

The Grenfell Tower disaster was the product of a systems failure compounded by the under-performance or failure of individual components. For example:

- the failure of flat 16 to contain the fire: '[T]he passage of fire across the face of the cladding and through the building was ... a result of inadequate barriers and compartmentation. ... [A]t Grenfell ... some construction details around the windows raise important questions about the adequacy of fire barriers and stops at each floor and junction. ... When it comes to fire stopping, attention to detail is all-important, as a small breach can prove to be fatal' (Gorse and Sturges 2017: 74–75);
- the failure of the aluminium cladding to resist the fire. It burned: '[T]he external facing materials, including the cladding, combusted too easily' (Gorse and Sturges 2017: 72);
- the failure of the aluminium cladding to prevent the fire leap-frogging from one level to the next. More effective fire breaks within the cladding may have prevented this.

The Grenfell Tower disaster offers important lessons for regulators, architects, engineers, manufacturers, constructors and clients. It also raises questions. For example:

- are high-rise residential buildings, by virtue of their complexity, coupling, density and size, an *inherently* unsafe technology?;
- is it possible to retrofit high-rise buildings without compromising their engineering integrity/increasing their risk-loading? As demonstrated by the Nimrod loss (see above), retrofits can introduce new failure modes and increase risk-loading, especially if they are undertaken without due regard for system integrity and resilience and fail to consider the possibility of unanticipated interactions between new or modified components/systems and the original engineered system. In the case of Nimrod XV230, it is believed that an element of the Supplementary Cooling Pack (SCP) – a retrofitted system – ignited spilled fuel. High-rise buildings should not be retrofitted with inferior-quality components/systems. High-rise buildings should not be modified in such a way that new failure modes are introduced. At Grenfell, the cladding created a chimney around the outside of the building. Worse, the chimney itself was flammable. Reactive patching (see Weir's (1996) work) increases risk-loading. As demonstrated by the loss of Nimrod XV230 and the Piper Alpha and

Grenfell Tower disasters, reactive patching with inferior-quality and/or poorly designed components/systems invites catastrophe.

CULTURE

Culture, as in 'the way we do things *here*' (Helmreich and Merritt 2001: 1), created the conditions that led to disaster. Specifically:

- Britain's embracing of modernism after the Second World War saw terraces replaced with high-rises and corner shops replaced with shopping arcades. Even the Ronan Point disaster could not dissuade the Royal Borough of Kensington and Chelsea from building high in North Kensington. Once the ideology of modernism had embedded itself it was unassailable. Ideology is durable. It is sticky (hard to shift);
- the Thatcher revolution, perpetuated by the Blair and Brown governments, saw the hollowing-out of local government and local accountability, with the setting up of arms-length agencies such as tenant management organisations (TMOs) to manage council-owned properties;
- under Thatcher, Blair and Brown, social housing was framed not as a national asset but as a problem to be solved. Nye Bevan's utopian vision for social housing was effectively smashed. It is likely that the problematisation of social housing and stigmatisation of council tenants persuaded some councils that investing in either was a waste of money. It has been claimed that the RBKC installed Arconic's flammable product against residents' wishes: 'Documents submitted to Kensington and Chelsea Council's planning department show residents were consulted in 2012 over the renovations and were asked what cladding they wanted. They show they chose a fire-resistant product called VMZ Composite which was said to have "many benefits". A newsletter handed to tenants and submitted with the planning application stated: "Various cladding options have been shown to residents with the composite cladding system being favoured by the majority". The document clearly stated the cladding had "fire retardancy". Two years later, a cheaper scheme was agreed and new proposals were approved by council planners. Instead of the fire-resistant panels chosen by residents, cheaper plastic-filled cladding

was fitted' (Tubb 2017). In her post-Grenfell report *Building a Safer Future*, Hackitt (2018: 11) observed: '[T]he voices of residents often [go] unheard, even when safety issues are identified';

- given that lessons were not learned from fires in other high-rise blocks, it is reasonable to conclude that neither the regulator, nor the various parties involved in commissioning, designing, constructing, maintaining and refurbishing the RBKC's high-rise residential blocks, possessed a learning culture. Gorse and Sturges (2017: 72) put it succinctly: 'The industry has become blind, and in some cases has ignored the lessons which should have been learned'. An organisation that possesses a learning culture 'is committed to learn safety lessons; communicates them to colleagues; remembers them over time' (Carthey and Clarke 2010: 10). The Transportation Safety Board of Canada (2014a: 7) observes: 'An organisation with a strong safety culture is generally proactive when it comes to addressing safety issues'. Lessons could have been learned from the 2009 Lakanal House fire and from several high-rise fires in Dubai: 'In a series of recent high-rise fires in Dubai, there occurred rapid fire spread up vertical faces through the polyurethane and aluminium cladding. Examples include the 82-storey Torch Tower [2015], the 75-storey Sulafa Tower [2016], and the 63-storey The Address Downtown Dubai Tower [2016]. ... In The Address building, the fire was presumed to have started with an electrical fault that spread through the window construction to the cladding, similar to Grenfell' (Gorse and Sturges 2017: 74). These episodes created opportunities for active learning (see Toft (1992) and Toft and Reynolds (1997) for a definition). With regards to the RBKC's Grenfell Tower, there was none;
- a regulatory culture of muddling-through undermined the coherence and comprehensibility of the building regulations. Gorse and Sturges (2017: 72) observe: '[A] review of the [building] regulations is at best confusing, and some parts of the Approved Documents are misleading. ... [I]t is doubtful whether the parties were aware of what standards they should be building to. ... [T]he regulations do not provide an intelligible message'. Hackitt (2018: 11) observes: '[T]he package of regulations and guidance (in the form of Approved Documents) can be ambiguous and inconsistent'.

PROCESSES/PROCEDURES

Flawed processes and procedures helped create the conditions for disaster. For example:

- the fashion for desk-top studies meant that some components were not tested physically;
- a building could be declared fit for purpose without any understanding of its systemic properties: 'Individual elements are being used as part of compound systems that are not being fully tested as systems' (Hackitt 2017: 18). Only if the systemic or synergistic properties of a system are understood can one be certain that every failure mode has been identified;
- the tenant voice, if heard, could be dismissed or ignored;
- there was little active learning from incidents, accidents and near-misses. Fires in high-rises in the UK, Middle East and Australia held important lessons. In 2014, a fire at the Lacrosse building in Melbourne that 'raced up 13 floors … in 11 minutes, was blamed on flammable aluminium composite cladding that lined the exterior concrete walls' (Wahlquist 2017).

Conclusions

The Grenfell Tower disaster gestated over many decades. The seeds were sown:

- in the aftermath of war, when central and local government embraced modernist solutions to the housing crisis;
- with councillors' and town planners' promotion of massive slum-clearance projects;
- with councillors' and town planners' debunking of alternative solutions to the housing crisis, such as the rehabilitation of tired Victorian and Edwardian properties;

- when clients, architects and engineers decided not to apply the principle of redundancy to key building elements – which led to many tower-blocks, including Grenfell, being built with a single staircase;
- when regulators, clients, architects and engineers decided that a residential tower-block's safety system need not include sprinklers. Defence-in-depth sacrificed on the altar of financial conservatism?;
- when Conservative and Labour governments began problematising council housing and council tenants. Why throw money at a housing sector that is in managed decline? During the 5 November 2018 Guy Fawkes night celebration, the hosts of a bonfire party held in the back garden of a house in Norwood, London, set light to a cardboard model of Grenfell Tower, then posted a video: 'The film, apparently shot ... in a back garden where there is a George cross flag in the background, shows cut-outs of faces at the windows. Brown paper had been used for most of them' (Dalton 2018b). Prime Minister Theresa May visited the Grenfell disaster scene on 15 June. Although she talked with emergency services personnel, she did not talk with survivors or victims' relatives. When society's leaders set a poor example, ugly things happen. Indifference spawns indifference;
- when it was decided to install cheaper thermal cladding that was less resistant to flame.

Conclusions

Systems-thinking explains the origins of incident, accident and near-miss. By revealing the back-story it offers a counterpoint to the false certainties of reductionism and injustices of blamism. It is inherently moral.

The concepts that support systems-thinking, such as coupling, emergence, practical drift, satisficing and reactive patching, promote understanding and create opportunities for active learning. However, as demonstrated by the loss of Nimrod XV230 and the Grenfell Tower disaster, such opportunities are not always taken. Regarding the loss of Nimrod XV230, early warnings included the loss of a Tornado 'where fuel had been drawn into the lagging round a hot pipe, and had ignited' (Maffett 2008) and the rupture of a Supplementary Cooling Pack (SCP) duct in Nimrod XV227. Regarding the Grenfell fire, early warnings included several high-profile fires in the Middle East and a fire in a London high-rise.

Incidents, accidents and near-misses often exhibit near-identical failure modes. For example, the Piper Alpha disaster, Nimrod loss and Grenfell Tower fire were triggered by reactive patching. In the case of the Piper Alpha disaster, by Occidental's decision to add a gas-processing unit. In the case of the Nimrod loss, by the MoD's decision to add an in-flight refuelling capability, supplementary cooling packs and other items. According to Coroner Andrew Walker, upgrades so increased the ageing airframe's interactive complexity that it became unsafe. Walker observed: 'I am satisfied that the design modifications to the [Nimrod] made the aircraft unsafe to fly' (Walker cited in Maffett 2008). The Coroner based his conclusion on a review of 'a mass of impenetrable documentation' pertaining to the aircraft's iterations (Maffett 2008). In the case of the Grenfell Tower fire, by the local authority's decision to add thermal cladding. Concepts such as reactive patching help foreground decisions and actions that invite or provoke incident, accident and near-miss.

Systems-thinking reveals the politics of failure. Often, disasters originate in politicians', civil servants', regulators', designers' and manufacturers' historic decisions. Such decisions may have more to do with fad and fashion – zeitgeist, if you will – than rationality. The political roots of Grenfell include Britain's post-war love-affair with modernism, disdain for conservationism (at least until the late 1970s), the hollowing-out of local government, residualisation, tolerance of a confused and confusing building code, tolerance of dubious practices such as desk-top estimates of the resilience of building materials and system elements and the problematic relationship between Grenfell's tenants and the Kensington and Chelsea Tenant Management Organisation. Given that the Blair and Brown governments tolerated the hollowing-out of local government, the Labour Party's post-Grenfell attacks on the May government smack of hypocrisy.

Systems-thinking reveals the economics of failure. The economic roots of Grenfell include the decision not to provide the block with two staircases, the decision not to install a building-wide fire alarm system, the decision not to retrofit a sprinkler system and the decision not to install a more fire-resistant cladding.

Through the medium of actor-network theory (ANT) we can understand how prime-movers organise for success. An actor-network is an assemblage of like-minded individuals, supportive institutions, helpful technologies and useful narratives that survives and thrives by recruiting new members and turning opponents. Members (actants) cohere through shared beliefs and common purpose. Network dysfunctions include over-commitment (groupthink).

Actor-network theory renders the origins of incidents, accidents and near-misses visible. While the Fukushima nuclear disaster was triggered by a sub-sea quake, it was incubated by Japan's pro-nuclear lobby – an actor-network consisting of politicians, civil servants, industrialists, shareholders, trade unions, academics, sympathetic journalists, nuclear technology, a culture of risk-taking and a determination to develop a reliable supply of energy borne of Japan's historic vulnerabilities. Similarly, while the Lac-Mégantic derailment and fire was triggered by an engineer not setting sufficient brakes on his train, the disaster was incubated by the Canadian government's determination to quickly exploit an economic opportunity. To this end, it created an actor-network consisting of politicians, civil servants, rail companies, lobbyists, track, locomotives, tank

Conclusions

cars, the politics of arms-length regulation and a national reconstruction narrative.

Actor-networks are expressions of ambition and power. An actor-network underwritten by a persuasive and popular ideology (such as national reconstruction) can achieve much. Of course, the ends served are not necessarily moral. During the Second World War, the Nazis created an actor-network of collaborators, extermination camps, track, locomotives, cattle trucks, pseudo-science, propagandists, the politics of Aryanism and the politics of anti-Semitism in pursuit of the Final Solution (Endlösung). The actor-network of the Final Solution proved highly effective, dispatching some 6 million Jews. Other victims of the Nazi actor-network-of-death included the Roma, Jehovah's Witnesses, left-wing politicians and activists, homosexuals and the mentally infirm.

Because it exposes the back-story, systems-thinking has the potential to safeguard against victimisation and, *in extremis*, miscarriages of justice. As demonstrated by the Munich air disaster and Lac-Mégantic rail disaster, those most directly involved can find themselves bearing the entire burden of blame. In the aftermath of the Munich air disaster, Captain Thain was criticised for failing to identify a threat (degraded take-off performance occasioned by runway contamination) about which little was known. The vilification was indiscriminate: Thain's daughter Sebuda was bullied at school on account of her father's alleged recklessness. The episode broke him.

Following the Lac-Mégantic rail disaster, engineer Tom Harding, traffic controller Richard Labrie and manager of train operations Jean Demaitre were portrayed as the villains of the piece. They were handcuffed, taken to prison and charged with forty-seven counts of criminal negligence. In Canada, the charge of criminal negligence causing death carries a maximum penalty of life imprisonment. Harding's arrest was theatrical: 'Mr Harding was surrounded at his home by a SWAT team and led away in handcuffs … despite the fact that his lawyer had notified the police that Mr Harding would voluntarily come to the court. … He was escorted to a makeshift courtroom in full view of the news media' (Quigley 2017).

In January 2018, after nine days of deliberation in the context of the TSBC's meticulous, systems-thinking-informed analysis of the

Lac-Mégantic disaster, a jury acquitted the three MMA employees on all counts. The TSBC's documenting of the back-story helped the court make an informed decision. It helped ensure that right was done.

Of course, like all methodologies, systems-thinking has its imperfections and imponderables. It provokes difficult questions. For example, is it reasonable, in the aftermath of a disaster, to question decisions taken years earlier by actors who acted in good faith? Is it reasonable to question decisions taken by actors under circumstances specific to that time? Is it reasonable to question decisions taken by actors on the basis of data that, subsequently, is shown to be partial, flawed or speculative? Is it reasonable to question decisions taken by actors whose advisors are subsequently shown to be incompetent, unreliable or treacherous?

By limiting the breadth or depth of an investigation one risks missing important detail. The wider the purview, the more accurate the analysis. The Nazi state committed heinous acts. However, the Nazis were, to a degree, a product of the punitive reparations imposed on Germany after the Great War and of the French occupation in 1923 of the Ruhr. Triumphalism and indifference breed resentment. Poverty, hopelessness and shame lend demagoguery respectability.

In 1986, the Space Shuttle Challenger disintegrated shortly after take-off, killing, amongst others, astronaut Sharon Christa McAuliffe, selected from over 11,000 applicants to be the first NASA Teacher in Space. Given the visibility and political import of the Teacher in Space programme – US President Ronald Reagan had taken a keen interest in the initiative – NASA was under pressure to meet its deadlines and get McAuliffe into space. On balance of probabilities, the visibility and politics of the mission influenced NASA's decision to launch in unfavourable temperatures. Decision-making is context-sensitive. To quote Monteiro (again): 'You do not go about doing your business in a total vacuum, but rather under the influence of a wide range of ... factors' (Monteiro 2012). The White House was culpable in the disaster.

Actions must be interpreted thoughtfully. To a degree, everyone is a prisoner of circumstance and history. English law is informed by the principle *fiat justitia* – Let right be done. Systems-thinking can help ensure this high ideal is realised in the matter of investigating incident, accident and near-miss.

Glossary of terms

Active learning: A form of learning where lessons learned from incident, accident and near-miss inspire safety improvements (see Toft (1992) and Toft and Reynolds (1997) for a definition).

Bimodality: A way of describing a component or system (for example, a light-bulb) whose performance-range is either: working (functioning); or non-working (non-functioning). In terms of function, a light-bulb does not have a degraded mode.

Emergence: Due to unanticipated interactions, a situation where a complex system behaves in unexpected ways or produces unexpected results (as when a computer programme produces an unexpected instruction or answer). Behaviour is described as emergent when a system 'exhibits behaviours that cannot be identified through functional decomposition' (Johnson 2005: 1). 'Emergent outcomes are ... not predictable from knowledge of their components, and not decomposable into those components' say Hollnagel, Wears and Braithwaite (2015: 38).

Groupthink: A process whereby the members of a tight-knit, under-pressure work-group see the world, and interpret intelligence, in the same way. Mind-guards keep order. Dissenters are ostracised or excluded and out-groups undermined. Members claim moral superiority and consider themselves invulnerable (Janis 1972; Neck and Moorhead 1995).

High-fidelity investigation: A form of incident or accident investigation that recognises the contribution of systemic factors. A high-fidelity investigation is inclusive/holistic/panoptic.

High-reliability organisation: High-reliability organisations (HROs) demonstrate: a preoccupation with failure (HROs learn from incidents, accidents and near-misses and ensure that remediations work); a reluctance to simplify (every incident, accident and near-miss is investigated); sensitivity to operations (staff appreciate the interactive complexity of socio-technical systems and are aware of phenomena like satisficing, practical drift and safety migration); resilience (HROs create capacity to meet unexpected demands and resist shocks); deference to expertise (HROs value problem-solving ability) (Roberts 1990; LaPorte and Consolini 1991; Christianson, Sutcliff, Miller and Iwashyna 2011). Practical HRO tools include 'reliability professionals' (mid-level personnel who promote safety) and communication frameworks such as the situation-task-intent-concern-calibrate (STICC) rule used by US firefighters to co-ordinate effort (Christianson, Sutcliff, Miller and Iwashyna 2011). Proponents claim HRO theory can improve safety in systems-of-systems (mega-systems). High-reliability organisation theory challenges Normal Accident Theory (NAT).

Interactive complexity: The possibility that system components will interact (that is, act on one another) in unexpected (that is, unanticipated) ways: 'Interactive complexity refers to component interactions that are non-linear, unfamiliar, unexpected or unplanned, and either not visible, or not immediately comprehensible for people running the system' (Woods et al. 2010: 62).

Intractability: Intractability obtains when 'it is difficult or impossible to follow and understand how [systems] function' (Hollnagel, Wears and Braithwaite 2015: 38). Intractability is opacity.

Latent threat/error/failure: According to Maurino, Reason, Johnston and Lee (1998: 13) 'latent failures … are mainly associated with weaknesses in, or absence of defences'. According to Helmreich (2000: 783), latent threats are 'existing conditions that may interact with ongoing activities to precipitate error'. For example: a fatigued pilot is less able to execute a difficult approach; An oil production platform lacking blast walls is more vulnerable; A flight-crew that has been stressed by a disruptive passenger may under-perform during an in-flight emergency.

Liveware: The human component in a socio-technical system or system-of-systems.

Loose coupling: A loosely coupled system is one with slack or 'give' (Perrow 1984). Loosely coupled systems are multi-channel systems. Loose coupling nurtures resilience. During the Vietnam War, President Richard M. Nixon attempted to disrupt the flow of men and materiel between North and South Vietnam by bombing Cambodia. At best, American B52 strikes reduced the volume of supplies transiting Cambodia by 10 per cent. By driving myriad trails through Cambodia, the Communists had created a loosely coupled, multi-channel system of supply that, by virtue of its engineered resilience, was more or less immune to bombing: '[I]nterdiction could not work against an enemy who moved his goods on human backs, along foot or bike trails' (Ambrose 1985: 243).

Mindfulness: A state of 'enriched awareness … that facilitates the … discovery and correction of unexpected events capable of escalation' (Weick, Sutcliffe and Obstfeld 1999: 37). To be mindful is to be sensitised to threat.

Multimodality: A way of describing a component or system where performance varies: 'Humans and organisations are … multimodal, in the sense that their performance is variable – sometimes better and sometimes worse, but never failing completely. A human "component" cannot stop functioning and be replaced in the same way as a technological component can' (Hollnagel, Wears and Braithwaite 2015: 37).

Non-linear interactions: A phenomenon where a system functions in unexpected ways to produce unexpected results: 'Non-linear interactions – where large inputs generate unexpectedly small outputs, small inputs generate unexpectedly large outputs and, through time, identical inputs generate qualitatively different outputs – render the behaviour of complex, output-oriented systems unpredictable' (Bennett 2016: 153).

Normal accident: Perrow (1984) posits a positive relationship between complexity and failure. As technologies become more complex, the potential for unanticipated interactions, both within the system, and between the system and its environment, grows. Failure to manage unanticipated interactions (using either hard or soft defences) may lead to disaster: 'A normal accident is where everyone tries very hard to play safe, but unexpected interaction of two or more failures (because of interactive complexity), causes a cascade of failures (because of tight coupling). The combination of complexity and coupling will bring the system down despite all safety efforts' (Perrow 2012). The pessimism of NAT contrasts with the optimism of high-reliability organisation (HRO) theory.

Normalisation of deviance: A process whereby, over time and in the context of adequate performance, deviations from prescribed standards and practices are normalised (Vaughan 1996). The normalisation of deviance is especially problematic in risky activities such as aviation (Bennett 2009) and space exploration (Vaughan 1996).

Opacity: A system attribute. An opaque system is one whose workings are invisible or poorly understood. In their best-selling novel *Fail-Safe*, Burdick and Wheeler (1962) explored the risks inherent in the superpowers' adoption of high-speed, opaque electronic military command-and-control devices. In the 1964, Sidney Lumet-directed motion picture of the same name, a system malfunction leads to a squadron of Convair B-58 Hustler nuclear-armed bombers being dispatched to Moscow.

Passive learning: A situation where learning opportunities do not spur action(s).

Practical drift: A situation where, to get the job done, pressured staff (all grades) ignore procedure(s). The concatenation of production pressures with resource shortages may cause staff to ignore rules and regulations in order to meet targets. Staff frame the drift away from prescribed practice as 'practical' (pragmatic) because it serves a higher purpose (Snook 2000; Bennett 2009).

Reactive patching: The modification of a socio-technical system in response to endogenous perturbations (for example, the wearing-out of a component, requirement to decommission a capability or incorporate a new capability) and exogenous perturbations (for example, the need to accommodate a skills shortage): 'All complex sociotechnical systems tend to operate in degraded mode. Under normal operating conditions, the actual state of the system will usually contain improvements, short-cuts, error-correcting routines, and other elements "patched" into the system. ... *Many of these "patches" will not be documented* [my emphasis]' (Weir 1996: 116).

Regulatory capture: Regulatory capture obtains when an element of a socio-technical system (for example, a corporation) is able to exert undue influence over a regulator: 'Regulatory capture happens when a regulatory agency, formed to act in the public's interest, eventually acts in ways that benefit the industry it is supposed to be regulating, rather than the public' (Investopedia 2018). Stigler (1971: 3) notes: 'Regulation may be actively sought by an industry, or it may be thrust upon it. ... [A]s a rule, regulation is acquired by the industry and is designed and operated primarily for its benefit'.

Resilience: The capacity of a system to resist endogenous or exogenous perturbation(s) to maintain normal working. Perturbations may be human or non-human in origin (for example, employee sabotage, a stock-market collapse, a shortage of raw materials or adverse weather).

Resilience-engineering: The process of organising for resilience. That is, organising an activity or designing a device with the aim of maximising its resilience. Devotees of HRO use elaboration to build resilience: specifically, the duplication or triplication of key hardware and live-ware. Devotees of NAT use simplification to build resilience: specifically, the elimination of components to affect a reduction in interactive complexity, intractability and opacity. Hopkins (1999: 99) is sceptical of HRO: '[R]edundancy often causes accidents: it increases interactive complexity and opaqueness and encourages risk taking'.

Safety I and Safety II: Hollnagel's typology distinguishes between safety methodologies that focus on why things go wrong (Safety I) and safety methodologies that focus on why things go right (Safety II). Hollnagel, Wears and Braithwaite (2015: 4) attribute safe operation to employees' knowledge, intuition, invention and adaptability: 'Things do not go right because people behave as they are supposed to, but because people can and do adjust what they do to match the conditions. ... As systems ... introduce more complexity, these adjustments become increasingly important. ... The challenge for safety improvement is ... to understand these adjustments – in other words, to understand how performance usually goes right in spite of the uncertainties, ambiguities, and goal conflicts that pervade complex work situations. Despite the obvious importance of things going right, traditional safety management has paid little attention to this. ... Safety-II ... relates to the system's ability to succeed under varying conditions. A Safety-II approach assumes that ... performance variability provides the adaptations that are needed to respond to varying conditions, and hence is the reason why things go right. Humans are consequently seen as a resource necessary for system ... resilience'.

Safety migration: The tendency for systems to operate/be operated at the outer limits of the safety envelope (Rasmussen 1997, 1999). Employees who interpret the absence of failure as proof that failure is impossible may be tempted to take risks. Production pressure may induce risk-taking: 'Rasmussen has argued that front-line workers do not follow strict procedures but follow the most useful and productive path. ... Workers operate within an envelope of possible actions and are influenced by organisational and social forces. Violations become more frequent and serious with time, so that the whole system "migrates" towards the boundaries of safety' (de Saint Maurice et al. 2010: 328).

Satisficing: A process whereby persons engaged in an activity settle for adequate rather than optimal results. While such behaviour might reflect a lack of commitment or application, it might also be provoked by adverse operating conditions (such as fatiguing rosters or unrealistic production targets).

Socio-technical system: Any system that combines people and things (technological artefacts) to achieve some predetermined goal (such as the provision of air service, manufacture of cars, ships or aircraft or building of residential towers). The term comes from Trist and Bamforth's (1951) studies of the organisation of men and machinery for the purpose of mining coal. In 1972, Professor Elwyn Edwards broke new ground with his SHEL model, a representation of aviation as a dynamic socio-technical system. Edwards's original model was modified by Hawkins (1993) who added a second 'L' to represent human-human and human-machine interaction (see Figure 23).

System accident: An accident that originates in the architecture and *modus operandi* of a system. System accidents often feature interactive complexity, tight coupling, emergence, intractability, practical drift and other systemic phenomena. A system accident involves 'a sequence of unanticipated compounding malfunctions and breakdowns' (Deepwater Horizon Study Group 2011: 10).

System-as-designed: A blueprint-informed description of a socio-technical system.

System-as-found: A description of a socio-technical system based on the system's current incarnation (including deletions, additions, modifications and work-arounds).

Systems-thinking: A way of thinking about problems such as criminality, social breakdown, technological or corporate failure that references both

Glossary of terms

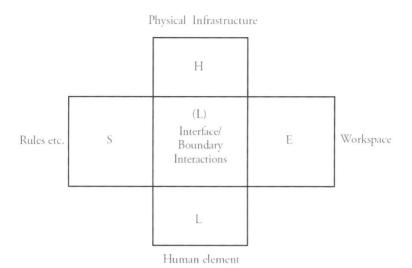

Figure 23. Edwards's innovative SHEL(L) model of aviation as a socio-technical system.

immediate and proximate causes (such as materialism, consumerism, design error and economic perturbations).

Tight coupling: A tightly coupled system is one with little slack or 'give'. A tightly coupled production process is invariant. It has little slack, has a single mode and is run to a schedule (Perrow 1984). As discussed above, in his 1964 motion picture *Fail-Safe*, Hollywood director Sidney Lumet explored the consequences for world peace of relying on high-speed, tightly coupled, opaque computer programmes for the command and control of advanced military hardware, such as the United States' fleet of nuclear-armed bombers. Lumet's picture tells the story of how an unforeseen computer malfunction leads to the dispatch of a fleet of Convair B-58 nuclear-armed bombers to Moscow (see Figure 24). *Fail-Safe* ponders the consequences for safety of tight coupling in hardware, software *and* live-ware – Lumet uses a coruscating scene in which the wife of one of the pilots tries unsuccessfully to convince him that he is the victim of a false alarm to make the point that tightly coupled (that is, inflexible) live-ware is as much a threat to safety as tightly coupled hardware and software: '[Lumet directs a] pointed critique ... at the rigidity of military protocols that preclude the pilots carrying out the nuclear attack from exercising reason or judgment in response to pleas from family members who explain that their mission was mistakenly initiated. Instead, trained, obedient soldiers become an extension of a faulty technological apparatus, literally executing the programs of an attack scenario designed to prevent any deviation from orders, no matter how illogical or apocalyptic they may be' (Anderson 2013). Coupling at operator-level (which can lead to mindlessness (Weick, Sutcliffe and Obstfeld 1999)) compounds problems associated with coupling at machine-level.

Work-as-imagined: According to Hollnagel, Wears and Braithwaite (2015: 40–41), 'What designers, managers, regulators and authorities believe happens, or should happen'. Sometimes, management instructions have unexpected consequences because they are premised on structures that no longer exists or practices that have been modified or abandoned.

Figure 24. In the Cold War motion-picture classic *Fail-Safe*, a malfunction in a tightly coupled, high-speed command-and-control computer sees a fleet of Convair B-58s tasked to destroy Moscow (Wikimedia Commons 2018n).

Bibliography

Agerholm, H. (2017). 'Grenfell Tower memorial: Survivors and victims' families still struggle to come to terms with fatal tragedy'. <http://www.independent.co.uk/news/uk/home-news/grenfell-tower-memorial-fire-families-victims-survivors-dead-killed-st-pauls-kensington-a8111181.html> accessed 9 January 2018.

Ambrose, S. E. (1985). *Rise to Globalism. American foreign policy since 1938*. Harmondsworth: Penguin Books.

American Bureau of Shipping (2005). *Guidance Notes on the Investigation of Marine Incidents*. Houston, TX: American Bureau of Shipping.

Anderson, S. (2013). 'Chaos and Control'. <http://scalar.usc.edu/anvc/chaosandcontrol/the-extensions-of-computer-fail-safe-and-dr-strangelove> accessed 22 November 2018.

Associated Press (2016). '7 killed in China's latest coal mine blast, bringing week's total mining deaths to 60. Blast in Hubei province follows death of 53 miners in two similar accidents last week'. <https://www.scmp.com/news/china/economy/article/2052110/11-trapped-chinas-latest-coal-mine-explosion> accessed 10 October 2018.

Australian Transport Safety Bureau (2007). *Regional Airline Line Operations Safety Audit. Report B2004/0237*. Canberra: Australian Transport Safety Bureau.

Baker, J. (2007). *The Report of the BP U.S. Refineries Independent Safety Review Panel*. Washington, DC: BP U.S. Refineries Independent Safety Review Panel.

Beck, M., and Drennan, L. (2000). *Offshore Risk Management: Myths, Worker Experiences and Reality*. Paper presented at the Qualitative Evidence-based Practice Conference, Coventry University, 15–17 May 2000.

Bell, J. (2006). *The Causes of Major Hazard Incidents and How to Improve Risk Control and Health and Safety Management: A Review of the Existing Literature*. Buxton: Health and Safety Laboratory.

Bennett, S. A. (2009). 'Aviation Security: At Risk from Practical Drift?' In P. Seidenstat and F. X. Splane (eds), *Protecting Airline Passengers in the Age of Terrorism*, pp. 33–62. Santa Barbara, CA: Praeger Security International.

Bennett S. A. (2010). 'Human Factors for Maintenance Engineers and Others – A Prerequisite for Success'. In R. Blockley and W. Shyy (eds), *Encyclopaedia of Aerospace Engineering*, pp. 4703–4710. Chichester: Wiley.

Bennett, S. A. (2016). 'Disasters and Mishaps: The Merits of Taking a Global View'. In A. J. Masys (ed.), *Disaster Forensics: understanding root cause and complex causality*, pp. 151–173. Cham: Springer International Publishing.

Bennhold, K. (2017). 'Ferraris and Opera Were Urgent, but Grenfell Tower Risks Went Unheeded', *The New York Times*, 15 August.

Bolton, G. (2010). *Reflective Practice, Writing and Professional Development*. 3rd edn. Thousand Oaks, CA: Sage.

Bomey, N. (2016). 'BP's Deepwater Horizon costs total $62 billion', *USA Today*, 14 July.

Booth, R., Bowcott, O., and Davies, C. (2018). 'Expert lists litany of serious safety breaches at Grenfell Tower', *The Guardian*, 4 June.

British Broadcasting Corporation (2010). 'Timeline: BP oil spill'. <http://www.bbc.co.uk/news/world-us-canada-10656239> accessed 30 March 2018.

British Broadcasting Corporation (2018a). 'Dozens missing in deadly Russia explosion in Magnitogorsk'. <https://www.bbc.co.uk/news/world-europe-46720260> accessed 1 January 2019.

British Broadcasting Corporation (2018b). 'On this day 1950–2005. 1957: Inquiry publishes cause of nuclear fire'. <http://news.bbc.co.uk/onthisday/hi/dates/stories/november/8/newsid_3181000/3181342.stm> accessed 9 February 2018.

British Petroleum (2010). *Deepwater Horizon Accident Investigation Report, September 8 2010*. London: British Petroleum.

Brunette, A. (2018). 'New railway track to bypass downtown Lac-Mégantic', *Canadian Broadcasting Corporation News*, 8 May.

Bulman, M. (2017). 'Grenfell Tower graphic: what we know about how the fire spread. Blaze spread up and across the building within minutes', *The Independent*, 15 June.

Burdekin, S. (2003). 'Mission Operations Safety Audits (MOSA): Measurements of behavioural Performance or non-Technical Skills from Military Aircrew'. *Aviation Safety Spotlight*, 0403.

Burdekin, S. (2015). 'Mission Operations Safety Audit. 36SQN pilot performance evaluation using MOSA'. *Aviation Safety Spotlight*, 03.

Burdick, E., and Wheeler, H. (1962). *Fail-Safe*. New York, NY: McGraw-Hill.

Burnham, D. (1970). 'Graft Paid to Police Here Said to Run Into Millions', *The New York Times*, 25 April.

Callon, M., and Latour, B. (1981). 'Unscrewing the Big Leviathan: How Actors Macro-Structure Reality and How Sociologists Help Them To Do So'. In K. Knorr-Cetina and A. V. Cicourel (eds), *Advances in Social Theory and Methodology: Towards an Integration of Micro and Macro-Sociology*, pp. 277–303. London: Routledge.

Callon, M., and Law, J. (1997). 'After the Individual in Society: Lessons on Collectivity from Science, Technology and Society', *Canadian Journal of Sociology*, 22(2), 165–182.

Canadian Broadcasting Corporation News (2013). 'Head of railway in Lac-Mégantic disaster says he's "also a victim"', *Canadian Broadcasting Corporation News*, 27 December.

Campbell, B. (2013a). 'Lac-Mégantic: where does the buck stop?' <https://www.straight.com/news/516511/bruce-campbell-lac-megantic-where-does-buck-stop> accessed 26 October 2018.

Campbell, B. (2013b). *The Lac-Mégantic Disaster. Where Does the Buck Stop?* Ottawa: Canadian Centre for Policy Alternatives.

Carthey, J., and Clarke, J. (2010). *Implementing Human Factors in Healthcare*. London: Department of Health.

Cerny, P. G. (2010). 'Globalisation and Statehood'. In M. Beeson and M. Bisley (eds), *Issues in 21st Century World Politics*, pp. 17–32. Basingstoke: Palgrave-Macmillan.

Challenger, R., Clegg, C. W., and Robinson, M. (2009). *Understanding Crowd Behaviours: Guidance and Lessons Identified*. London: Cabinet Office.

Christianson, M. K., Sutcliff, K. M., Miller, M. A., and Iwashyna, T. J. (2011). 'Becoming a high reliability organisation', *Critical Care*, 15(314), 1–5.

Columbia University (2009). 'Japan's Modern History: An Outline of the Period.' <http://afe.easia.columbia.edu/timelines/japan_modern_timeline.htm> accessed 27 December 2016.

Crouch, C. (2017). *Can Neoliberalism Be Saved From Itself?* London: Social Europe.

Dalton, D. (2018a). 'Chernobyl Shelter To Begin Full Operation In December, Says Ukraine President', *NucNet*, 27 April.

Dalton, J. (2018b). 'Laughing party-goers burn model of Grenfell Tower in "callous and sickening" bonfire night scenes. Model – reported to police as hate crime – had brown faces at windows', *The Independent*, 5 November.

Davis, M. C., Challenger, R., Jayewardene, D. N. W., and Clegg, C. W. (2014). 'Advancing socio-technical systems thinking: A call for bravery', *Applied Ergonomics*, 45(2A), 171–180.

Day, P. (2013). 'Aberdeen's central role in the North Sea oil industry'. <https://bbc.co.uk/news/> accessed 9 November 2018.

De Laurentis, D. (2005). 'Understanding Transportation as System-of-System Design Problem', 43rd AIAA Aerospace Sciences Meeting and Exhibition, Reno, NV.

de Saint Maurice, G., Auroy, Y., Vincent, C., and Amalberti, R. (2010). 'The natural lifespan of a safety policy: violations and system migration in anesthesia', *Quality and Safety in Health Care*, 19, 327–331.

Deepwater Horizon Study Group (2011). *Final Report on the Investigation of the Macondo Well Blowout*. University of California, Berkeley, CA: Centre for Catastrophic Risk Management, Department of Civil and Environmental Engineering.

Dekker, S. W. A. (2006). 'Resilience Engineering: Chronicling the Emergence of Confused Consensus'. In E. Hollnagel, D. D. Woods and N. Leveson (eds), *Resilience Engineering: Concepts and Precepts*, pp. 77–92. Aldershot: Ashgate Publishing Ltd.

Dekker, S. W. A. (2014a). 'The bureaucratisation of safety', *Safety Science*, 70, 348–357.
Dekker, S. W. A. (2014b). *The Field Guide to Understanding 'Human Error' (Third Ed.)*. Farnham: Ashgate Publishing Ltd.
Department of the Navy (2017). *Comprehensive Review of Recent Surface Force Incidents*. Norfolk, VA: Department of the Navy.
Devlin, A. (2018). 'What is the Grenfell Tower Inquiry, who is Sir Martin Moore-Bick and when's the report on the fire going to be published?' *The Sun*, 3 October.
Dewey, J. (1933). *How We Think. A restatement of the relation of reflective thinking to the educative process*. Revised edition. Boston, MA: D. C. Heath.
Disraeli, B. (1845). *Sybil or The Two Nations*. London: Henry Colburn.
Dörner, D. (1996). *The Logic of Failure: Recognising and Avoiding Error in Complex Situations*. Cambridge, MA: Perseus Books.
Duncan, J. (2017). 'RIBA responds to Grenfell Tower Inquiry Terms of Reference', *Press Release*, 17 August.
The Economist (2017). 'Britain: back to being the sick man of Europe?' <https://www.economist.com/buttonwoods-notebook/2017/07/19/britain-back-to-being-the-sick-man-of-europe> accessed 30 September 2018.
Edwards, E. (1972). 'Man and machine: systems for safety'. In *Proceedings of British Air Line Pilots Association Technical Symposium*, pp. 21–36. London: British Air Line Pilots Association.
Erlanger, S. (2017). 'After Grenfell Tower Fire, U.K. Asks: Has Deregulation Gone Too Far?' *The New York Times*, 28 June.
Fairlie, H. (1955). 'Political Commentary', *The Spectator*, 23 September.
Farnsworth, C. H. (1992). 'Canadian Judge Calls Air Crash Avoidable', *The New York Times*, 27 March.
Federal Aviation Administration (2018). 'Line Operations Safety Assessments (LOSA)'. <https://www.faa.gov/about/initiatives/maintenance_hf/losa/> accessed 20 April 2018.
Finding Petroleum (2013). 'Review: Lord Cullen – what have we learned from Piper Alpha?' <http://www.findingpetroleum.com/n/Review_Lord_Cullen_what_have_we_learned_from_Piper_Alpha/044b5113.aspx> accessed 29 March 2018.
Flight Safety Foundation Editorial Staff (2008). 'Line Operations Safety Audit (LOSA) Provides Data on Threats and Errors', *Focus on Commercial Aviation Safety*, Summer, 17–23.
Fricker, J. (1975). 'The RAF. Lean yet potent'. In *Royal Air Force Yearbook 1975*, pp. 2–14. London: The Royal Air Force Benevolent Fund Enterprises.
Gall, C. (2010). '40 years of North Sea oil: Life on the rigs in the early days was like the wild west', *Daily Record*, 7 October.
Gapper, J. (2017). 'Grenfell: an anatomy of a housing disaster'. <https://www.ft.com/content/5381b5d2-5c1c-11e7-9bc8-8055f264aa8b> accessed 11 October 2018.

Glendon, A. I., Clarke, S. G., and McKenna, E. F. (2006). *Human Safety and Risk Management*. Boca Raton, FL: CRC Press.

Gorse, C., and Sturges, J. (2017). 'Not what anyone wanted: Observations on regulations, standards, quality and experience in the wake of Grenfell', *Construction Research and Innovation*, 8(3), 72–75.

Green, W. (1975). *The Observer's Book of Aircraft*. London: Frederick Warne and Co. Ltd.

Grotan, T. O. (2013). 'Assessing risks in systems operating in complex and dynamic environments'. In E. Albrechtsen and D. Besnard (eds), *Oil and Gas Technology and Humans. Assessing the Human Factors of Technological Change*. Aldershot: Ashgate Publishing Ltd.

The Guardian (2013). 'Piper Alpha disaster: how 167 oil rig workers died'. <https://www.theguardian.com/business/2013/jul/04/piper-alpha-disaster-167-oil-rig> accessed 30 September 2018.

Hackitt, J. (2017). *Building a Safer Future. Independent Review of Building Regulations and Fire Safety: Interim Report*. London: Her Majesty's Stationery Office.

Hackitt, J. (2018). *Building a Safer Future. Independent Review of Building Regulations and Fire Safety: Final Report*. London: Her Majesty's Stationery Office.

Haddon-Cave, C. (2009). *The Nimrod Review. An Independent Review into the Broader Issues Surrounding the Loss of the RAF Nimrod MR2 Aircraft XV230 in Afghanistan in 2006. HC 1025*. London: Her Majesty's Stationery Office.

Hall, A. (1995). *Emergency Rescue*. Enderby: Bookmart Ltd.

Hall, S., and Jacques, M. (1989). 'Manifesto for New Times'. In S. Hall and M. Jacques (eds), *New Times. The changing face of politics in the 1990s*, pp. 23–37. London: Lawrence and Wishart.

Hall, S., and Jacques, M. (eds) (1989). *New Times. The changing face of politics in the 1990s*. London: Lawrence and Wishart.

Harris, D. (2014). 'Improving aircraft safety', *The Psychologist*, 27(2), 90–94.

Hawkins, F. H. (1993). *Human Factors in Flight*. Aldershot: Ashgate Publishing Ltd.

Health and Safety Executive (2017). *Offshore Statistics and Regulatory Activity Report 2016*. Liverpool: Health and Safety Executive.

Helmreich, R. L. (2000). 'On error management: lessons from aviation', *British Medical Journal*, 320, 781–785.

Helmreich, R. L., and Merritt, A. (2001). *Culture At Work in Aviation and Medicine*. Aldershot: Ashgate Publishing Ltd.

Henry, C. (2007). 'The Normal Operations Safety Survey (NOSS): Measuring System Performance in Air Traffic Control Safety Systems', Second Institution of Engineering and Technology International Conference on System Safety, 22–24 October, Savoy Place, London.

Holden, R. J. (2009). 'People or systems? To blame is human. The fix is to engineer', *Professional Safety*, 54(12), 34–41.

Hollnagel, E. (2004). *Barriers and Accident Prevention*. Aldershot: Ashgate Publishing Ltd.

Hollnagel, E. (2016). The ETTO Principle – Efficiency-Thoroughness Trade-Off. <http://erikhollnagel.com/ideas/etto-principle/index.html> accessed 31 March 2018.

Hollnagel, E., and Leonhardt, J. (2013). *From Safety-I to Safety-II: A White Paper*. Brussels: European Organisation for the Safety of Air Navigation.

Hollnagel E., Wears, R. L., and Braithwaite, J. (2015). *From Safety-I to Safety-II: A White Paper*. Copenhagen: University of Southern Denmark.

Homer, A. W. (2009) 'Coal Mine Safety Regulation in China and the USA', *Journal of Contemporary Asia*, 39(3), 424–439.

Hopkins, A. (1999). 'The limits of normal accident theory', *Safety Science*, 32, 93–102.

Horton, R. (2017). 'Offline: Racism – the pathology we choose to ignore', *The Lancet*, 390, 14.

House of Commons Science and Technology Committee (2012). *Devil's bargain? Energy risks and the public. First Report of Session 2012–2013*. London: House of Commons Science and Technology Committee.

Hughes, D. (2017). 'Grenfell Tower fire: Council leader claims sprinklers were not fitted as residents did not want prolonged disruption'. <https://www.independent.co.uk/news/uk/home-news/grenfell-tower-fire-latest-sprinklers-not-fitted-residents-nick-paget-brown-claim-a7792736.html> accessed 12 October 2018.

Institute of Medicine (2000). *To Err Is Human. Building a Safer Health System*. Washington, DC: Institute of Medicine/National Academy of Medicine.

International Atomic Energy Agency (2015). *The Fukushima Daiichi Accident: Report by the Director General*. Vienna: International Atomic Energy Agency.

International Civil Aviation Organisation (1995). 'Six years after the Dryden tragedy, many accident investigation authorities have learned its lessons', *International Civil Aviation Organisation Journal*, 50(8), 20–25.

International Civil Aviation Organisation (2002). *Line Operations Safety Audit (LOSA)*. Montreal: International Civil Aviation Organisation.

International Maritime Organisation (2008). *Adoption of the Code of the International Standards and Recommended Practices for a Safety Investigation into a Marine Casualty or Marine Incident (Casualty Investigation Code)*. London: International Maritime Organisation.

Investopedia (2018). 'Regulatory Capture'. <https://www.investopedia.com/terms/r/regulatory-capture.asp> accessed 13 February 2018.

Jack, I. (1996). *Before the Oil Ran Out. Britain 1978–1986*. London: Vintage.

Jack, I. (2013). 'North Sea oil fuelled the 80s boom, but it was, and remains, strangely invisible', *The Guardian*, 19 April.

Janis, I. L. (1972). *Victims of Groupthink: A Psychological Study of Foreign-Policy Decisions and Fiascos*. Boston, MA: Houghton Mifflin.

Johnson, C. W. (2005). *What are Emergent Properties and How Do They Affect the Engineering of Complex Systems?* Glasgow: Department of Computing Science, University of Glasgow.

Jones, J. KBE (2017). *'The patronising disposition of unaccountable power' A report to ensure the pain and suffering of the Hillsborough families is not repeated*. London: The Stationery Office.

Khoshkhoo, R. (2017). 'Adaptation of Line Operations Safety Audit (LOSA) to Dispatch Operations (DOSA)', *Journal of Airline and Airport Management*, 7(2), 126–135.

Kiernan, R. (2016). 'Surface tension – the battle against the North Sea oil bosses', *Socialist Worker*, 29 July.

Kilgannon, C. (2010). 'Serpico on Serpico', *The New York Times*, 22 January.

Kingston, J. (2012). 'Japan's Nuclear Village', *The Asia-Pacific Journal*, 10(37). <http://apjjf.org/2012/10/37/Jeff-Kingston/3822/article.html> accessed 18 January 2017.

Klinect, J. R., Wilhelm, J. A., and Helmreich, R. L. (1999). 'Threat and error management: Data from line operations safety audits'. In *Proceedings of the Tenth International Symposium on Aviation Psychology*, pp. 638–688, May, Columbus, OH.

Lagadec, P. (1982). *Major Technical Risk: An Assessment of Industrial Disasters*. Oxford: Pergamon Press.

LaPorte, T. R., and Consolini, P. (1991). 'Working in practice but not in theory: theoretical challenges of high-reliability organisations', *Journal of Public Administration Research and Theory*, 1, 19–47.

Law, J. (ed.) (1991). *A sociology of monsters: essays on power, technology and domination*. London: Routledge.

Lawrence, M. (2018). 'Hackitt Review: "Need for a radical rethink" on fire safety'. <https://www.24housing.co.uk/news/hackitt-review-branded-disappointing-for-not-going-far-enough> accessed 11 October 2018.

Leroux, M. (2008). 'Captain James Thain cleared of blame after the thawing of hostilities', *The Times*, 30 January.

Ma, M., and Rankin, W. (2012). *Implementation Guideline for Maintenance Line Operations Safety Assessment (M-LOSA) and Ramp LOSA (R-LOSA) Programmes, DOT/FAA/AM-12/9*. Washington, DC: Office of Aerospace Medicine, Federal Aviation Administration.

Maas, P. (1973). *Serpico. The cop who defied the system*. New York, NY: Viking Press.

Macalister, T. (2011). 'Background: What caused the 1970s oil price shock?' *The Guardian*, 3 March.

Macalister, T. (2013). 'Piper Alpha Disaster: how 167 oil rig workers died', *The Guardian*, 4 July.
McDonald, A., Garrigan, B., and Kanse, L. (2006). 'Confidential observations of rail safety (CORS): An adaptation of line operations safety audit'. In *Proceedings of the Swinburne University Multimodal Symposium on Safety Management and Human Factors*. 9–10 February, Melbourne, Australia.
McIntyre, G. R. (2000). *Patterns in Safety Thinking*. Aldershot: Ashgate Publishing Ltd.
McKee, M. (2017). 'Grenfell Tower fire: why we cannot ignore the political determinants of health. A public health response must confront the underlying causes', *British Medical Journal*, 357.
McKenzie, B. (2015). 'Who Drives to Work? Commuting by Automobile in the United States: 2013'. <https://www.census.gov/library/publications/2015/acs/acs-32.html> accessed 7 March 2018.
Mackrael, K., and Robertson, G. (2014). 'Lax safety practices blamed for Lac-Mégantic tragedy', *Globe and Mail*, 19 August.
MacLeod, C. (2014). 'China mine disasters point to poor safety record', *USA Today*, 20 August.
Maffett, S. (2008). 'Nimrod design faults highlighted'. <http://news.bbc.co.uk/1/hi/uk/7417384.stm> accessed 9 November 2018.
Maier, M. W. (1998). 'Architecting principles for systems of systems', *Systems Engineering*, 1(4), 267–284.
Manchester Evening News (2008). 'Pilot's daughter pays tribute', *Manchester Evening News*, 5 February.
Mason, R. O. (2004). 'Lessons in Organisational Ethics from the Columbia Disaster: Can a Culture be Lethal?' *Organisational Dynamics*, 33(2), 128–142.
Maurino, D. E., Reason, J. T., Johnston, N., and Lee, R. (1998). *Beyond Aviation Human Factors*. Aldershot: Ashgate Publishing Ltd.
Mendick, R. (2017). 'Chimney effect: Grenfell's unusual design led blaze to spread, say investigators', *The Daily Telegraph*, 25 June.
Merton, R. K. (1936). 'The Unanticipated Consequences of Purposive Social Action', *American Sociological Review*, 1(6), 894–904.
Miller, K. D. (2009). 'Organisational Risk after Modernism', *Organisation Studies*, 30(2–3), 157–180.
Ming-Xiao, W., Tao, Z., Miao-Rong, X., Bin, Z., and Ming-Qiu, J. (2011). 'Analysis of National Coal-mining Accident Data in China, 2001–2008', *Public Health Reports*, 126(2), 270–275.
Mogford, J. (2005). *Fatal Accident Investigation Report. Isomerisation Unit Explosion Final Report, Texas City, Texas, USA*. Naperville, IL: BP Products North America Inc.

Möller, S. (2017). 'London's Grenfell Tower Fire and the Financialisation of Local Governance'. <http://www.governancexborders.com> accessed 1 May 2018.

Monteiro, E. (2012). 'Actor-network theory and information infrastructure'. <http://www.idi.ntnu.no/~eriem/ant.FINAL.htm> accessed 1 March 2016.

Moore-Bick, M. (2017). 'Grenfell Tower Inquiry – Terms of Reference'. Letter. 10 August 2017.

Morgan, K. O. (1987). *Labour People*. Oxford: Oxford University Press.

Moshansky, V. P. (1992). *Moshansky, Commission of Inquiry into the Air Ontario Accident at Dryden, Ontario: Final Report (Volumes 1–4)*. Ottawa: Minister of Supply and Services.

Murray, R. (1989). 'Fordism and Post-Fordism'. In S. Hall and M. Jacques (eds), *New Times. The changing face of politics in the 1990s*, pp. 38–53. London: Lawrence and Wishart.

National Aeronautics and Space Administration (2013). *The Case for Safety: The North Sea Piper Alpha Disaster*. Washington, DC: National Aeronautics and Space Administration.

The National Diet of Japan (2012). *The official report of the Fukushima Nuclear Accident Independent Investigation Commission. Executive Summary*. Tokyo: The National Diet of Japan.

Neck, C. P., and Moorhead, G. (1995). 'Groupthink Remodelled: The Importance of Leadership, Time-pressure and Methodical Decision-Making Procedures', *Human Relations*, 48(5), 537–557.

The Observer (2015). 'The vast gap between rich and poor in our capital is a crisis for us all'. <https://www.theguardian.com/commentisfree/2015/mar/08/observer-view-on-london> accessed 11 January 2018.

Officer of the Watch (2013). 'Alexander L. Kielland Platform Capsize Accident – Investigation Report'. <https://officerofthewatch.com/2013/04/29/alexander-l-kielland-platform-capsize-accident> accessed 22 March 2018.

Osborne, S., and Agerholm, H. (2018). 'Grenfell Tower inquiry: Refurbishment turned building into "death trap using public funds"', *The Independent*, 5 June.

Pasha-Robinson, L. (2017). 'Grenfell Tower inquiry ignoring systemic problems in "betrayal" of victims, warn MPs'. <http://www.independent.co.uk/news/uk/home-news/grenfell-tower-inquiry-mps-terms-reference-betrayal-downing-street-sir-martin-moor-bick-kensington-a7894701.html> accessed 9 January 2018.

Paterson, J. (1997). *Behind the mask. Regulating health and safety in Britain's offshore oil and gas industry. PhD Thesis (unpublished)*. Florence: European University Institute, Department of Law.

Perrow, C. (1983). 'The organisational context of human-factors engineering', *Administrative Science Quarterly*, 28(4), 521–541.

Perrow, C. (1984). *Normal Accidents: Living with High-Risk Technologies*. New York, NY: Basic Books.

Perrow, C. (2012). 'Getting to Catastrophe. Concentrations, complexity and coupling', *The Montréal Review*, December.

Perryman, N. (2006). *Fifties Britain: Post-War Life*. London: Bounty Books.

Pitcher, G. (2017). 'Javid admits possibility of "systemic failure" in Building Regs', *The Architects Journal*, 1 September.

Pollmann, M. (2016). 'Japan: How energy security shapes foreign policy'. <http://thediplomat.com> accessed 26 December 2016.

Quigley, K. (2017). 'A Lac-Mégantic trial, but where's the inquiry?' <https://beta.theglobeandmail.com/opinion/a-lac-megantic-trial-but-wheres-the-inquiry/article36220081/> accessed 6 November 2018.

Rasmussen, J. (1997). 'Risk management in a dynamic society: a modelling problem', *Safety Science*, 27, 183–213.

Rasmussen, J. (1999). 'The concept of human error: is it useful for the design of safe systems?' *Safety Science Monitor*, 3.

Reason, J. T. (1990). *Human Error*. Cambridge: Cambridge University Press.

Reason, J. T. (1997). *Managing the risks of organisational accidents*. Aldershot: Ashgate Publishing Ltd.

Reason, J. T. (2013). *A Life in Error*. Farnham: Ashgate Publishing Ltd.

Renn, O. (2008). 'Concepts of Risk: An Interdisciplinary Review. Part 1: Disciplinary Risk Concepts', *GAIA – Ecological Perspectives for Science and Society*, 17(1), 50–66.

Roberts, K. H. (1990). 'Some characteristics of one type of high-reliability organisation', *Organisation Science*, 1(2), 160–176.

Royal Academy of Engineering (2005). *Accidents and Agenda: Full sector reports. An examination of the processes that follow from accidents or incidents of high potential in several industries and their effectiveness in preventing further accidents*. London: Royal Academy of Engineering.

Rousseau, D. (1990). 'Quantitative assessment of organisational culture: The case for multiple measures'. In B. Schneider (ed.), *Frontiers in Industrial and Organisational Psychology, Volume 3*, pp. 153–192. San Francisco, CA: Jossey-Bass.

Schön, D. A. (1973). *Beyond the Stable State. Public and private learning in a changing society*. Harmondsworth: Penguin.

Schön, D. A. (1983). *The Reflective Practitioner. How Professionals Think In Action*. New York, NY: Basic Books.

Shorrock, S., Leonhardt, J., Licu, T., and Peters, C. (2014). *Systems Thinking for Safety: Ten Principles*. Brussels: Eurocontrol.

Single European Sky ATM Research unit (2018). 'SHELL Model'. <https://ext.eurocontrol.int/ehp/?q=node/1565> accessed 24 September 2018.

Snook, S. (2000). *Friendly Fire: The Accidental Shootdown of U.S. Black Hawks over Northern Iraq*. Princeton, NJ: Princeton University Press.
Stern, E. (2008). 'Crisis and Learning: A Conceptual Balance Sheet'. In A. Boin (ed.), *Crisis Management Volume III*. Los Angeles, CA: Sage.
Stigler, G. J. (1971). 'The Theory of Economic Regulation', *The Bell Journal of Economics and Management Science*, 2(1), 3–21.
Tadros, W. (2014). 'Release of Railway Investigation Report R13D0054'. <http://www.tsb.gc.ca/eng/medias-media/discours-speeches/2014/08/20140819.asp> accessed 24 October 2018.
Titov, V. V., and Synolakis, C. E. (1997). 'Extreme inundation flows during the Hokkaido-Nansei-Oki tsunami', *Geophys. Res. Lett*, 24(11), 1315–1318.
Todd, M. A., and Thomas, M. J.W. (2012). 'Flight Hours and Flight Crew Performance in Commercial Aviation', *Aviation, Space and Environmental Medicine*, 83, 776–782.
Toft, B. (1992). 'The Failure of Hindsight', *Disaster Prevention and Management: An International Journal*, 1(3), 48–63.
Toft, B., and Reynolds, S. (1997). *Learning from Disasters*. Leicester: Perpetuity Press.
Transport Canada (2014). *Transportation in Canada 2013. Overview Report*. Ottawa: Transport Canada.
Transportation Safety Board of Canada (2014a). *Lac-Mégantic runaway train and derailment investigation summary*. Gatineau: Transportation Safety Board of Canada.
Transportation Safety Board of Canada (2014b). *Runaway and main-track derailment. Montreal, Maine and Atlantic Railway freight train MMA-002, Mile 0.23, Sherbrooke subdivision, Lac-Mégantic, Quebec, 06 July, 2013*. Gatineau: Transportation Safety Board of Canada.
Trist, E., and Bamforth, K. (1951). 'Some social and psychological consequences of the longwall method of coal getting', *Human Relations*, 4(1), 3–38.
Tubb, G. (2017). 'Grenfell Tower residents were promised fire-resistant cladding five years ago', *Sky News*, 20 June.
Tucker, P. (2017). 'The Grenfell Tower fire was the end result of a disdainful housing policy', *The Guardian*, 20 June.
Turner, B. A. (1978). *Man-Made Disasters (First Ed.)*. London: Wykeham Publications.
Turner, B. A. (1994). 'Causes of Disaster, sloppy management', *British Journal of Management*, 5(3), 215–219.
Vaughan, D. (1996). *The Challenger Launch Decision. Risky Technology, Culture and Deviance at NASA*. Chicago: Chicago University Press.
Walker, P. (2013). 'Lakanal House tower block fire: deaths "could have been prevented". Deaths of six people in UK's worst tower block fire could have been prevented by proper fire-safety checks, inquest concludes', *The Guardian*, 28 March.

Wahlquist, C. (2017). 'Cladding in London high-rise fire also blamed for 2014 Melbourne blaze'. <https://www.theguardian.com/uk-news/2017/jun/15/cladding-in-2014-melbourne-high-rise-blaze-also-used-in-grenfell-tower> accessed 30 October 2018.

Wei, L. J., Hu, J. K., Luo, X. R., and Liang, W. (2017). 'Study and analyse the development of China coal mine safety management [*sic*]', *International Journal of Energy Sector Management*, 11(1), 80–90.

Weick, K. E., and Sutcliffe, K. M. (2007). *Managing the Unexpected: Resilient Performance in an Age of Uncertainty*. San Francisco, CA: Jossey-Bass.

Weick, K. E., Sutcliffe, K. M., and Obstfeld, D. (1999). 'Organising for high reliability: processes of collective mindfulness', *Research in Organisational Behaviour*, 21, 81–123.

Weir, D. T.H. (1996). 'Risk and disaster: the role of communications breakdown in plane crashes and business failure'. In C. Hood and D. K.C. Jones (eds), *Debates in Risk Management*, pp. 114–126. London: UCL Press.

Wenger, E. (1998). *Communities of Practice: Learning, Meaning and Identity*. Cambridge: Cambridge University Press.

Whyte, D. (2006). 'Regulating Safety, Regulating Profit: cost cutting, injury and death in the North Sea after Piper Alpha'. In E. Tucker (ed.), *Working Disasters: the politics of recognition and response*, pp. 181–207. New York: Baywood.

Wikimedia Commons (2018a). 'Entrance to a small coal mine in China, 1999'. <https://upload.wikimedia.org/wikipedia/commons/2/29/Sidings_and_shaft_entry.jpg> accessed 9 October 2018.

Wikimedia Commons (2018b). 'RAF Hawker Siddeley Nimrod XV230 at the 2005 Waddington Air Show'. <https://upload.wikimedia.org/wikipedia/commons/7/77/Nimrod_Waddington_2005_2.jpg> accessed 25 April 2018.

Wikimedia Commons (2018c). 'USS John S. McCain'. <https://commons.wikimedia.org/wiki/File:USS_John_S_McCain_South_China_Sea_1.JPG> accessed 10 November 2018.

Wikimedia Commons (2018d). 'USS John S. McCain'. <https://commons.wikimedia.org/wiki/File:US_Navy_170821-N-OU129-022_Damage_to_the_portside_of_USS_John_S._McCain_(DDG_56).jpg> accessed 10 November 2018.

Wikimedia Commons (2018e). 'An aerial view of tsunami damage in Tōhoku'. <https://en.wikipedia.org/wiki/2011_T%C5%8Dhoku_earthquake_and_tsunami#/media/File:SH-60B_helicopter_flies_over_Sendai.jpg> accessed 25 April 2018.

Wikimedia Commons (2018f). 'Calder Hall, the world's first full-scale atomic power station, is here shown nearing completion'. <https://commons.wikimedia.org/wiki/Category:Windscale_Piles#/media/File:HD.15.003_(11824034284).jpg> accessed 25 April 2018.

Wikimedia Commons (2018g). 'IAEA experts depart Unit 4 of TEPCO's Fukushima Daiichi Nuclear Power Station on 17 April 2013 as part of a mission to review Japan's plans to decommission the facility'. <https://commons.wikimedia.org/wiki/Category:Fukushima_I_accidents#/media/File:IAEA_Experts_at_Fukushima_(02813336).jpg> accessed 24 April 2018.

Wikimedia Commons (2018h). 'Oil Rig, Cromarty, Scotland'. <https://commons.wikimedia.org/wiki/File:CromartyOilPlatform.JPG> accessed 4 October 2018.

Wikimedia Commons (2018i). '1971 Dodge Challenger R/T coupe'. <https://commons.wikimedia.org/wiki/File:1971_Dodge_Challenger_R-T_coupe_(5410358104).jpg> accessed 1 October 2018.

Wikimedia Commons (2018j). 'Location of Brent oil platform in the North Sea'. <https://commons.wikimedia.org/wiki/File:Brent_crude_oil_map.png> accessed 4 October 2018.

Wikimedia Commons (2018k). 'BP's Deepwater Horizon rig on fire in the Gulf of Mexico'. <https://commons.wikimedia.org/wiki/File:BP%E2%80%99s_Deepwater_Horizon_rig_on_fire_in_the_Gulf_of_Mexico._(15010938489).jpg> accessed 25 April 2018.

Wikimedia Commons (2018l). 'Grenfell Tower Fire'. <https://commons.wikimedia.org/wiki/File:Grenfell_Tower_fire_(wider_view).jpg> accessed 10 November 2018.

Wikimedia Commons (2018m). 'One of several views of Grenfell Tower, nearly 11 months after the fire, with scaffolding covering most of the building'. <https://commons.wikimedia.org/wiki/File:Grenfell_Tower_in_May_2018_08.jpg> accessed 10 November 2018.

Wikimedia Commons (2018n). 'United States Air Force Convair B-58A Hustler in flight (B-58A-15-CF, SN 60-1118)'. <https://upload.wikimedia.org/wikipedia/commons/3/35/B-58_Hustler.jpg> accessed 21 November 2018.

Williamson, C. (2017). 'This is how neoliberalism, led by Thatcher and Blair, is to blame for the Grenfell Tower disaster'. <http://www.independent.co.uk/voices/grenfell-tower-inquiry-deregulation-thatcher-tony-blair-fire-service-cuts-a7876346.html> accessed 10 January 2018.

Woods, D. D., Dekker, S., Cook, R., Johannesen, L., and Sarter, N. (2010). *Behind Human Error*. Farnham: Ashgate Publishing Ltd.

Woolgar, S., and Latour, B. (1986). *Laboratory life: the construction of scientific facts*. Princeton, NJ: Princeton University Press.

World Nuclear Association (2016). 'Fukushima Accident'. <http://www.worldnuclear.org/informationlibrary/safetyandsecurity/safetyofplants/fukushimaaccident.aspx> accessed 14 August 2016.

World Nuclear News (2012). 'New Japanese regulator takes over'. <http://www.world-nuclear-news.org> accessed 17 January 2017.

Xu, B., and Albert, E. (2017). 'Media Censorship in China'. <https://www.cfr.org/backgrounder/media-censorship-china> accessed 9 October 2018.

Yamamori, H., and Mito, T. (1993). 'Keeping CRM is keeping the flight safe'. In E. L. Wiener, B. G. Kanki and R. L. Helmreich (eds), *Cockpit Resource Management*, pp. 399–420. San Diego, CA: Academic Press.

Youle, E., Gray, J., Slawson, N., and White, N. (2018). 'An End To Austerity? This Is What People At The Sharp End Say Should Happen Now'. <https://www.huffingtonpost.co.uk/entry/end-of-austerity-people-at-sharp-end-of-society-this-is-what-we-think_uk/> accessed 11 October 2018.

Index

Figures and Tables are indicated by page numbers in bold print.

Abe, Shinzo 46
active learning 125
actor-network theory (ANT) 1, 9–17, 54, 122, 123
  actants 9–10
  counter-networks 14
  and Japanese nuclear programme 54
  translation 16
  versatility 14
Address Downtown Dubai Tower fire (2016) 118
adverse events **99**
African Union 11
Air Ontario 20, 22
Alnic MC 31
amakudari (Japanese cultural trait) 54
American Bureau of Shipping (ABS): *Guidance Notes on the Investigation of Marine Incidents* 29–30
Arab League 11
Australia: Queensland Rail 95–6
aviation safety 80
  Line Operations Safety Audit (LOSA) 88–94, 97
    domains 89
    methodology 91–2
    possible risks 93–4
    proactive risk management 90–1
    threat and error 90
    virtuous circle **93**

BEA Airspeed Ambasssador crash (1958) 2–3, 123
  investigation 2–3
Bell, J. 65–6, 68
Bennett, S. A. 2
Bevan, Nye 106, 117
Blair, Tony 102, 117
blame and cover-up 43–4, 121, 123
Booth, R. et al 114
BP Macondo Deepwater Horizon disaster (2010) 4, 77–88, **78**
  Blow Out Preventer (BOP) failure 82, 83
  BP report 79
  buildings and infrastructure 82–3
  culture 83–5
  Deepwater Horizon Study Group (DHSG) report 81, 86
  estimated cost 82
  goals 79–82
  people 85–7
  processes and procedures 83
  safety concerns and productivity 79–80, 84
  safety and lack of learning 84–5
  safety rhetoric and action 81
  safety and system 84, 85–6, 87, 88
  technology 82
BP Texas City disaster 4, 79, 80, 81, 84
  Baker report 85
Britain
  Buncefield explosion and fire 80–1, 84
  council tenants right to buy 105–6
  culture 55
  economic/social conditions 1970s 62–3, **69**, 70

Flixborough chemical plant disaster (1974) 76
Grenfell Tower fire (2017) 100–20, 121, 122
Hillsborough football stadium disaster 43–4
Lakanal House fire (2009) 109, 118
membership of European Economic Community 62
nuclear power and security 52
Piper Alpha oil-and-gas production platform accident (1988) 4, 59–77, 86
post-war economy 52, 75
RAF Nimrod loss (2006) 23, 25–8, 29, 116
Ronan Point explosion (1968) 110, 114, 115, 117
'sick man' of Europe 62, 70
Thatcher government 63
Windscale nuclear accident (1957) 46, 47–8
Brown, Gordan 117
Buncefield explosion and fire 80–1, 84
Burkhardt, Edward Arnold 37, 42
Burnham, David 13–14

Callon, Michelle 7
Campbell, B. 38, 41–2
Canada
  Dryden air disaster (1989) 19–23, 29
  Lac-Mégantic derailment and fire (2013) 34–44, 122–3, 123–4
Canada Transportation Act 41
Canadian Rail Operating Rules 40
Carthey, J. and Clarke, J. 71–2, 118
CCTV 15
Centre for European Reform 108
Challenger, R. et al 9, 45–6, 60, 79, 99, 105
Chernobyl nuclear accident 4, 44, 45, 46

Chinese deep-mining industry 14–16, 58
  actor-network 16
  deep-mining industry, translation 16
  entrance to mine **15**
  media reporting 15, 16
  reasons for poor safety 14–15
Clarke, Alan 63
Commonwealth of Independent States (CIS) 11
Confidential Observations of Rail Safety (CORS) **94–6**, 97
conformity 58
corporatism 54
crew error 2
Crouch, Colin 101–2
Cullen, Lord William 59, 72, 76–7, 80, 83, 107–8

Dekker, Sidney 5, 7, 84–5
Demaitre, Jean 123
depths of analyses in investigations **30**
Devlin, Amanda 103
Dickens, Charles 102
Disraeli, Benjamin 106–7
Dörner, Dietrich 7
Dounreay 44
Dryden air disaster (1989) 19–23, 29
  causes 21–3, **24**
  immediate 21–2
  proximate 22–3
  challenges of flight 20–1
  crash after takeoff 21
  investigation report 5–6, 19, 29
  passenger numbers 20
Dubai 118
Durk, David 13

Edwards, Elwyn 5, 7
emergence 8, 9, 125
environmental factors 46

Index

error, causes of 9
'Establishment' and power 8
European Union (EU) 11

factors in accidents/incidents 8–9
Fairlie, Henry 7–8
Federal Aviation Administration 89–90
fiat justitia principle 124
Flixborough chemical plant disaster (1974) 76
Fokker F28 airliner 20
Fourth Estate 58
Fukushima Daiichi nuclear accident (2011) 4, 44–58, 86
   back-up power systems 48–9
   buildings and infrastructure 50–1
   culture 53–5, 57
   defensiveness of civil servants 54
   earthquake and tsunami 44, **45**, 49, 50–1, 57
   Federation of Electric Power Companies (FEPC) 55
   Fukushima Daiichi nuclear accident (2011) 44–6, 48–58
   International Atomic Energy Agency experts **58**
   Japanese security and energy supplies 46, 51, 122
   National Diet report 54, 55, 57–8
   Nuclear and Industrial Safety Agency (NISA) 50, 51, 55
   nuclear power actor-network 54
   Nuclear Regulation Authority (NRA) 53
   Nuclear Safety Commission (NSC) 51–2
   people 55–7
   processes and procedures 51–3
   regulations 54–5
   systems-thinking investigation 45
   warnings 50–1

Gapper, J. 106, 107, 111, 112
Germany
   Munich BEA Airspeed Ambassador crash (1958) 2–3, 123
   pre-war culture 55
   reasons for rise of Nazis 124
   Second World War actor-network 123
Gorse, C. and Sturges, J. 109, 113, 114, 116, 118
Grenfell Tower fire (2017) **100**–20, 121, 122
   and austerity 109
   buildings/infrastructure 113–14
   charred hulk of Grenfell Tower **104**
   cladding 105, 113, 115, 116, 117–18, 121
   contracting-out 111–12
   contracting-out of safety roles 107
   culture 117–18
   diversity of residents 106
   early warnings 121
   fire: start and progress 113–14
   fire retardancy of building materials 107, 117
   ghettoisation 106
   Guy Fawkes Night burning of model 120
   Justice 4 Grenfell 108–9
   Kensington and Chelsea Tenant Management Organisation (KCTMO) 108–9, 111, 112
   learning culture 118
   lessons from disaster 116–17
   memorial service 100–1
   post-war council house and tower block building 110, 117, 119
   post-war modernism 117, 119, 122
   problematising of council housing and tenants 120
   processes/procedures 119
   public inquiry 101, 114

scope 101, 102
reducing public expenditure 109
residualisation in housing 106
Royal Borough of Kensington and Chelsea 111, 112, 117
seeds of disaster 119–20, 122
social/economic polarisation 102, 107, 112–13
sprinkler system 109
technology 114–17
　coupling/complexity 114–15
　failure of components 115–16
　retrofitting 116
　voices of residents unheard 118
Grotan, T. O. 84
groupthink 52, 54, 125

Hackitt, Dame Judith 103, 105, 107, 109, 110–11, 118
Haddon-Cave, Charles 23, 26, 29
Hall, S. and Jacques, M. 69
Harding, Tom 42, 43, 123
Helmreich, R. L. 90, 126
Helmreich, R. L. and Merritt, A. 22, 40, 53, 56, 117
high-reliability organisations 125–6
Hillsborough football stadium disaster 43–4
Holden, Richard 7
Hollnagel, Erik 5, 7
Hollnagel, Erik et al 67–8
Homer, A. W. 14, 15
hybrid-collectif 9, 11

immediate and proximate causes 29–30
imperfections of systems-thinking 124
informed culture 72
interactive complexity 126
International Atomic Energy Agency (IAEA) 45, 50, 54–5, **58**

International Civil Aviation Organisation (ICAO) 8, 9, 80, 89, 91
International Maritime Organisation (IMO) 29
International Monetary Fund (IMF) 11
intractability 7, 129, 130
iterations in systems-thinking 17

Jack, Ian 77
Japan
　Fukushima Daiichi nuclear accident (2011) 4, 44–58, 86
　mindfulness/mindlessness 52–**3**, 55–**6**
　pre-war culture 55
Johnson, Christopher 7
Jones, Rt. Rev. James 43–4
just culture 71

Kiernan, Raymie 74, 75
Kirkham, Francis 109
Klinect, J. R. et al 94, 97
Knapp Commission 13
Korea 56

Lac-Mégantic derailment and fire (2013) 34–44, 122–3, 123–4
　buildings and infrastructure 37–8
　Canadian austerity and cost cutting 42
　fire 35–6
　Montreal, Maine and Atlantic Railway Ltd. (MMA) 37, 38, 39–43
　people and blame 42–3
　political factors 41–2
　Transport Canada 38–9, 40–1
　TSBC criticism
　　cost cutting 36, 41, 42
　　crash-worthy standards 36
　　design and maintenance of equipment 36–7

Index

lax safety culture 39, 40–1
parking on a gradient 38
procedural failures 39–40
route planning 36, 38
rushed repair 35–6, 38
train crews 38
training programmes 36
unattended trains 38
Lagadec, P. 9, 60, 109
Lakanal House fire (2009) 109, 118
latent failures 126
Latour, Bruno 7
Law, John 7
learning culture 72
levels: micro and macro 10
Lindsay, John V. 13
Line Operations Safety Audit (LOSA)
 aviation 88–94, 97
 railways 94–7

McAuliffe, Sharon Christa 124
McDonald, A. et al 90, 94, 97
McKee, M. 108
Macmillan, Harold 47
Maintenance Aviation Safety Action Program (ASAP) 90
Manchester Evening News 3
marine accident investigations 29–34
Maurino, D. E. et al 6, 20, 23, 126
May, Theresa 109, 120
MD-11 crash 91
Mendick, R. 113, 114
Merton, R. K. 66
meta systems 10, 11–12
Miller, Kent D. 7
Mills, Keith 19–20, 21
mindful cultures 55, 58
mindfulness: definition 127
mindfulness/mindlessness 52–**3**, 55–**6**, 67
Ming-Xiao, W. et al 16
Möller, S. 108, 109

Monteiro, E. 10, 46, 88, 124
Montreal, Maine and Atlantic Railway Ltd. (MMA) 37, 38, 39–43
Moore-Bick, Sir Martin 101, 103
Moorwood, George 19–20, 21, 23
Moshansky, Virgil P. 5, 6, 19, 29
murahachibu (Japanese cultural trait) 53

National Aeronautics and Space Administration 65, 66
National Aeronautics and Space Administration Aviation Safety Reporting System (ASRS) 90
neoliberalism 63, 101
network space 9, **10**
New York Police Department (NYPD) corruption 12–14
New York Times 13–14
non-linear interactions 8, 127
normal accidents 127
normalisation of deviance 127–8
North American Free Trade Agreement (NAFTA) 11
North Sea oil and gas
 Alexander Kielland platform accident (1980) 60, 64
 oil and gas production accidents: death and injury 74, 75
 oil revenue 70
 Piper Alpha oil-and-gas production platform accident (1988) 59–77
 rigs in the Cromarty gas field 61
 safety cases 76
nuclear accidents 44–58
Nuclear Installations Act (1959) 47–8
nuclear power generation risks 49–50
nuclear power and national security 51–2

Obama, Barack 78
Occidental Petroleum 59, 64, 65, 66, 67, 68, 71, 76

oil and gas production
   in Canada and USA 40–1
   offshore accidents 59–88
   oil-shocks (1970s) **69**
   prices 68
opacity 128
open culture 71
Organisation of American States
   (OAS) 11
Organisation of the Petroleum Exporting
   Countries (OPEC) 11, 64, 75
organisations and assumptions/values 81
organisations and reputation 44
origins of systems thinking 7
   in aviation investigations 29

passive learning 128
Perrow, Charles 5, 7, 114
personal factors in accidents/incidents 9
Petrov, Stanislav 67–8
Piper Alpha oil-and-gas production
   platform accident (1988) 4,
   59–77, 86
   buildings and infrastructure 64–6
   causes 60, 68, 70–3, 75–6
   conversion for oil and gas
      production 64–5
   culture 68–73
   deluge system 66, 71
   economic pressures 64, 70
   eyewitness accounts 59–60, 61–2
   goals of North Sea oil and gas
      extraction 61–4
   loss of life and injury 59–60
   management 63, 66, 67, 70, 71, 72, 76
   Permission-to-Work system 68, 72
   processes and procedures 66–8
   safety concerns 61–2, 63–4, 65, 70, 76
   and risk-taking 75
   safety oversight 107–8
   technology 64

   workers' families 73
   gambling, sex and alcohol 73
   survival training 73
   trade unions 62
   value of 67–8, 71
   working conditions 61–2, 63–4,
      70–1, 73, 74
Principle of Generalised Symmetry 9
purposive social action: latent and
   manifest functions 66

RAF Nimrod loss (2006) 23, 25–8, 29,
   116
   contributing factors 26, 29
   early warnings 121
   Independent Review findings 23,
      25–6
   Nimrod Airworthiness Review
      Team (NART) 25
   Nimrod at Waddington Air Show **25**
   operational demands 27
   vulnerabilities revealed 27, **28**
Railway Association of Canada 41
railways: LOSA methodology and
   CORS **94–6**, 97
Rasmussen, J. 52, 64
Reagan, Ronald 124
Reason, James 5–6, 7, 40, 51, 67, 83
regulation 41–2
reliability of production systems 48
Renn, O. 76
Reporters Without Borders 16
reporting culture 72
Rousseau, D. 81
Russia: Magnitogorsk tower block
   explosion 115

safety audits 88–97
   aviation 88–94
   railways 94–7
safety culture **71–2**

safety and economic performance 80
Safety I and Safety II 67, 68, 129
safety management systems (SMSs) 41–2
safety migration 52, 129–30
safety slogans 80–1
safety and 'unconscious safety mind' 81
Sellafield 47
Serpico, Frank 12, 13
Single European Sky ATM Research unit (SESAR) 9
Smart, Gavin 112
Space Shuttle Challenger (1986) 124
Spectator, The 7
standards and recommended practices (SARPs) 29
Strauss, Lewis 52
Sulafa Tower 118
Sunday Times 58
supra-national actors 11
system accidents 6, 130
system behaviour and complexity 83
system characteristics 6–7
system-as-designed 8, 130
system-as-found 8, 130
system-of-systems 27
systems-thinking: defined 130–1

Tadros, Wendy 37, 41
Thain, James 2, 3, 43, 123
Thain, Sebuda 123
Thalidomide scandal 58
Thatcher, Margaret 63, 102, 105, 117
Titanic accident 60
Toft, B. 51
Tohoku undersea earthquake 44
Tokyo Electric Power Company (TEPCO) 44, 50, 51
Torch Tower 118
translation (actor-network theory) 16
Transport Canada 38–9, 40–1, 42
transportation of crude oil 41

Transportation Safety Board of Canada (TSBC) 35–6, 118
Turner, Barry 5, 7

Ukraine
  Chernobyl nuclear accident 4, 44, 45, 46
  New Safe Confinement (NSC) shelter 45
University of Queensland 94–7
University of Texas Human Factors Research Project 88
US Navy collisions and grounding (2017) 31–4
  Comprehensive Review of Recent Surface Force Incidents 32–3, 33–4
  immediate and proximate causes 32–3
  investigation 32
USA
  automobile culture and demand for hydrocarbons 87
  BP Macondo Deepwater Horizon disaster (2010) 4, 77–88, **78**
  BP Texas City disaster 4, 79, 80, 81, 84
  Navy collisions and grounding (2017) 31–4
  USS *Antietam* 31
  USS *Bellnap* 32
  USS *Fitzgerald* 31
  USS *John F. Kennedy* 32
  USS *John S. McCain* **31, 34**
  USS *Lake Champlain* 31

Vaughan, Diane 5, 7

Walker, Andrew 121
Wei, L. J. et al 16
Weick, K. E. et al 52, 67, 71, 91
Whyte, D. 63, 64, 69–70, 72

Williamson, Chris 102
Wilson, Harold 26
Windscale nuclear accident (1957) 46, **47**–8
Wood Group 74, 77

Woolgar, Stephen 7
workers as asset 67–8
World Bank 11
World Nuclear Association (WNA) 45, 50, 51